ALUMINIUM STRUCTURAL

Recent European Advances

ALUMINIUM STRUCTURAL ANALYSIS

Recent European Advances

A collection of papers on aspects of research and design in structural aluminium with particular reference to the needs of a future European code of practice.

Edited by

P. S. BULSON

ELSEVIER APPLIED SCIENCE
LONDON AND NEW YORK

ELSEVIER SCIENCE PUBLISHERS LTD
Crown House, Linton Road, Barking, Essex IG11 8JU, England

Sole Distributor in the USA and Canada
ELSEVIER SCIENCE PUBLISHING CO., INC.
655 Avenue of the Americas, New York, NY 10010, USA

WITH 48 TABLES AND 158 ILLUSTRATIONS

© 1992 ELSEVIER SCIENCE PUBLISHERS LTD

British Library Cataloguing in Publication Data

Aluminium structural analysis : recent European advances
I. Bulson, P. S. (Philip Stanley), *1925–*
624.1826

ISBN 1-85166-660-5

Library of Congress Cataloging-in-Publication Data

Aluminium structural analysis : recent European advances / edited by P.S. Bulson.
  p. cm.
  A collection of papers on aspects of research and design in structural aluminium with particular reference to the needs of a future European code of practice.
  Includes bibliographical references and index.
  ISBN 1-85166-660-5
  1. Aluminium construction. 2. Structural analysis. I. Bulson, P. S., 1925– .
TA690.A47 1991
624.1′826—dc20   91-31776
                                                              CIP

No responsibility is assumed by the Publisher for any injury and/or damage to persons or property as a matter of products liability, negligence or otherwise, or from any use or operation of any methods, products, instructions or ideas contained in the material herein.

Special regulations for readers in the USA

This publication has been registered with the Copyright Clearance Centre Inc. (CCC), Salem, Massachusetts. Information can be obtained from the CCC about conditions under which photocopies of parts of this publication may be made in the USA. All other copyright questions, including photocopying outside the USA, should be referred to the publisher.

All rights reserved. No part of this publication may be reproduced, stored in a retrieval system, or transmitted in any form or by any means, electronic, mechanical, photocopying, recording, or otherwise, without the prior written permission of the publisher.

Typeset and printed in Northern Ireland by The Universities Press (Belfast) Ltd.

Preface

This book looks ahead to the design of aluminium structures in the remaining years of this century and the early years of the next. It will be a time of change in the world of structural codes of practice and specifications for general engineering, with the introduction of Eurocodes and the need to harmonise these with national and international practice. Design methods are changing too, with most engineers now aware of limit state methods and their advantages.

There has been much interaction recently between the leading authorities in structural aluminium in Europe and North America, in anticipation of the transition to new codes of practice. In the UK the British Standards Institution is about to publish a new code on Structural Aluminium. In other countries of the European Community much effort is being put into the drafting of the aluminium sections of the European Convention for Constructional Steelwork codes. The Italian Standards organisation has produced a new Aluminium code of practice. The German government is examining a replacement for their longstanding code. There has been a recent resurgence of effort under the aegis of the International Standards Organisation towards an international code for Structural Aluminium and this, of course, has included the USA and Canada. Much needs to be done to effect a dovetailing of these diverse interests, all of which see the need for a progressive attitude to design.

The purpose of this book is to help towards an international view of the subject, with particular reference to the problem of bringing the

European authorities into a closely-knit mode. It has been decided, therefore, to bring together the writings of six acknowledged experts in the field from four European countries, each contributing to an aspect of the subject by drawing from long experience of structural aluminium in design, testing and analytical research. All the experts have made, or are making, major contributions to code-writing. This choice of authors is particularly important because the time is not too far distant when work on a new Eurocode on Aluminium Structural Design will be in full swing.

The book opens with a chapter by the editor and Dr Cullimore which is designed to set the scene by reviewing the development of aluminium in structural engineering, and which highlights design principles and problems. The design principles are aimed particularly at limit state design and design for reliability and economy. The problems include the influence on design of the low elastic modulus and its effect on stability, vibrations, fatigue and fracture mechanics; also the low melting point and the change in properties in heat-affected material. The chapter concludes with thoughts for the future.

The long second chapter is a major joint contribution by two well-known academics from the mainland of Europe. Professor F. Mazzolani from the Universita di Napoli and Professor G. Valtinat from the Technische University Hamburg-Harburg, FRG combine to discuss the behaviour of bars, beams, columns and beam-columns. There has been much fundamental research on this subject at their respective universities in support of the European Recommendations for Aluminium Alloy Structures (ERAAS), and Professor Mazzolani has published a text book on aluminium structures. Both professors are also members of the International Standards Organisation committees preparing an international standard for structural aluminium.

Although the second chapter includes reference to the torsional and lateral stability of members, it was thought appropriate that the work in these areas in the UK should be reviewed separately. Chapter three, therefore, is written by Professor Nethercot of the University of Nottingham, and summarises the considerable range of testing and analysis dealing with the torsional and lateral buckling of struts and beams, undertaken by specialists in British universities and elsewhere in the past 40 years. Professor Nethercot has contributed towards the preparation of the new UK code of practice, and is also concerned with the British input to the new International standard. Because he is also very well known in Europe for his work on steel structures, he is able to draw comparisons in the way the two materials are treated in design codes.

Professor H. R. Evans, head of the new School of Engineering at the University of Wales at Cardiff, is the leading expert on the design and analysis of aluminium plate girders. Research on plate girders has been associated with Cardiff for a long time, and the structural testing facilities there are of a high standard. Chapter four, therefore, gives Professor Evans an opportunity to summarise recent work at Cardiff on the ultimate strength of aluminium plate girders, and to show how this work has been developed for the new UK code of practice. It is hoped that this approach could form the basis of design clauses in a future aluminium Eurocode. The work at Cardiff has been aimed particularly at the effect on plate girder strength of welding, especially in higher strength alloys in the 6*** and 7*** series.

Problems of welded construction are also the subject of the following chapter, by Dr Soetens of TNO Building and Construction Research at Delft in the Netherlands. The stress analysis of welded joints and the influence of the heat affected zone is a particularly important subject, and Dr Soetens has become a leading authority in the field through his work at Delft. In recent publications he has examined the effect of deformation capacity on the behaviour of welded joints, and has used finite element analysis to simulate the behaviour to failure of typical welded test-specimens. Here he presents a state-of-the-art review of the analysis of fillet and butt-welded joints.

Dr Cullimore, formerly of Bristol University, writes on bolted and rivetted joints for general engineering structures in aluminium. He also summarises his own research into the strength of friction-grip bolted joints. His research on the analysis of the latter has been confirmed by extensive test programmes at Bristol, and his chapter is therefore a timely summary of recent progress. His work has been partially sponsored by the UK Ministry of Defence, whose military engineers have particular structural problems for which friction-grip bolted joints are the best answer.

We have brought together, therefore, a selection of experts from many of the major research areas in structural aluminium. Readers are reminded that the subject matter of the book is aimed at the design of aluminium buildings, bridges, ships, vehicles, towers and similar structures, but not at the design of aircraft, pressure vessels or space vehicles. There still exists a clear division between the general world of aluminium structures and the specialist world, such as aircraft structures, where safety, reliability and economy are bought at great expense. It will be interesting to see if these areas grow closer together as time goes by.

P. S. BULSON

Contents

Preface v
List of Contributors xi

1 DESIGN PRINCIPLES AND PROBLEMS 1
P. S. Bulson & M. S. G. Cullimore
1. INTRODUCTION 1
2. SAFETY AND SERVICEABILITY 5
3. HEAT-AFFECTED ZONES 7
4. TENSION MEMBERS 9
5. COMPRESSION MEMBERS 10
6. FATIGUE 13
7. FRACTURE MECHANICS 17
8. WELDED JOINTS 24
9. TESTING AND QUALITY ASSURANCE 25
10. CONCLUSIONS 28

2 BARS, BEAMS AND BEAM COLUMNS 35
F. M. Mazzolani & G. Valtinat
1. DEFINITION OF AN 'INDUSTRIAL BAR' 35
   1.1 General 35
   1.2 The Stress–Strain Relationship 36
   1.3 Geometrical Imperfections 43
   1.4 Mechanical Imperfections 47
2. MEMBERS IN TENSION 62
   2.1 General 62
   2.2 Strength of Elements 64
   2.3 Ductility of Connections 67
   2.4 Codification 69
3. MEMBERS IN BENDING 70
   3.1 General 70
   3.2 Ultimate Behaviour of Cross-Sections 75
   3.3 Plastic Behaviour of Statically Undetermined Girders 92
   3.4 Flexural Torsional Buckling 109
   3.5 Codification 127
4. MEMBERS IN COMPRESSION 131
   4.1 General 131
   4.2 Ultimate Behaviour of Cross-Sections 133
   4.3 Buckling of Columns 143
   4.4 Buckling of Beam Columns 169
   4.5 Codification 176

3 LATERAL-TORSIONAL BUCKLING OF BEAMS — 193
D. A. Nethercot

1. INTRODUCTION — 195
2. BASIS OF UK DESIGN PROCEDURE — 200
3. COMPARISON OF BS DESIGN CURVE AND THEORETICAL RESULTS — 202
4. EFFECT OF NON-UNIFORM MOMENT — 205
5. UNEQUAL FLANGED BEAMS — 209
6. TREATMENT OF SLENDER CROSS-SECTIONS — 211
7. CONCLUSIONS — 214

4 SHEAR WEBS AND PLATE GIRDERS — 219
H. R. Evans

1. INTRODUCTION — 219
2. GENERAL OBSERVATIONS ON THE NEW CODE — 221
3. UNSTIFFENED WEBS IN SHEAR (CLAUSE 5.6.2) — 223
4. TRANSVERSELY STIFFENED WEBS IN SHEAR (CLAUSE 5.6.3) — 226
5. LONGITUDINALLY STIFFENED WEBS IN SHEAR (CLAUSE 5.6.4) — 236
6. REQUIREMENTS FOR STIFFENERS (CLAUSE 5.6.5) — 239
7. LARGE OPENINGS IN WEBS (CLAUSE 5.6.6) — 243
8. GIRDERS UNDER COMBINED SHEAR AND BENDING (CLAUSE 5.6.7) — 244
9. COMPARISON OF BS 8118 AND CP 118 VALUES — 246
10. COMPARISON OF BS 8118 AND EXPERIMENTAL VALUES — 248
11. CONCLUSION — 251

5 WELDED CONNECTIONS — 253
F. Soetens

1. INTRODUCTION — 253
2. RESEARCH PROGRAMME FOR WELDED CONNECTIONS — 254
3. STATE OF THE ART — 255
4. EXPERIMENTAL RESEARCH ON MECHANICAL PROPERTIES — 264
5. EXPERIMENTAL RESEARCH ON FILLET WELDS — 279
6. EXPERIMENTAL AND THEORETICAL RESEARCH ON WELDED CONNECTIONS — 300
7. EVALUATION — 310

6 JOINTS WITH MECHANICAL FASTENERS — 313
M. S. G. Cullimore

1. INTRODUCTION — 315
2. SINGLE FASTENER JOINTS WITH IN-PLANE LOADING — 316
3. JOINTS WITH GROUPS OF FASTENERS — 321
4. PINNED JOINTS — 335
5. FRICTION GRIP BOLTED JOINTS — 345

Index — 367

List of Contributors

P. S. Bulson
Mott MacDonald Group, Advanced Mechanics and Engineering, 1 Huxley Road, Surrey Research Park, Guildford, Surrey, GU2 5RE, UK

M. S. G. Cullimore
Formerly at the Department of Civil Engineering, University of Bristol, UK. *Present address:* 1 Pitchcombe Gardens, Bristol, BS9 2RH, UK

H. R. Evans
School of Engineering, University of Wales, PO Box 917, Cardiff, CF1 3XH, UK

F. M. Mazzolani
Engineering Faculty, University of Naples, Piazzale Tecchio, 80125 Napoli, Italy

D. A. Nethercot
Department of Civil Engineering, University of Nottingham, University Park, Nottingham, NG7 2RD, UK

F. Soetens
TNO Building & Construction Research, Lange Kleiweg 5, Rijswijk, PO Box 49, 2600 AA Delft, The Netherlands

G. Valtinat
Technische Universitat, Hamburg-Harburg, Postfach 90 14 03, Laurenbruch Ost 1, 2100 Hamburg 90, Germany

1

Design Principles and Problems

P. S. Bulson

Mott MacDonald Group, Advanced Mechanics and Engineering, Guildford, Surrey, UK

&

M. S. G. Cullimore

Formerly University of Bristol, UK

ABSTRACT

After a short review of the growth of structural aluminium as a general engineering material, attention is focussed on some of the major design principles and problems faced by designers who use structural aluminium. Safety and serviceability are key elements of limit state design, and the use of partial factors of safety in recent codes of practice is discussed. A problem of particular interest to designers of welded aluminium structures is the presence of heat-affected zones and how these are allowed for in rules. The way that the new Code of Practice, BS 8118, deals with the contribution of heat-affected material to the strength of members is summarised.

The analysis of tension and compression members is discussed, and this is followed by a description of the difficulties associated with structural fatigue. Methods of dealing with this problem include the use of fracture mechanics for particular relationships between stress range and life. Design problems of welded joints are briefly mentioned, and the chapter concludes with a review of the problems of structural testing and quality assurance.

1 INTRODUCTION

Aluminium is an attractive material, light, strong and clean. It is not surprising that when first produced chemically it was classed as a precious metal, and it was only after the discovery of cheap methods of production from bauxite using electrolytic processes that its use in engineering structures became a possibility. Luckily the properties of the alloys of aluminium fitted the requirements of aircraft designers, so there was much money available for research and development from the 1920s onwards. The development of new alloys and of new methods of production was linked to the expansion of the military and civil aircraft industries and to the advent of the all-metal aircraft body. The need for new materials in the aeronautical and aerospace fields led the way for research in the past, and it still does today.

In the early days aircraft wing structures often used extruded aluminium alloy booms, or booms formed from shear webs and reinforced wing cover. Fuselages used thin sheeting reinforced with light stringers. Increases in ultimate tensile strengths were brought about by varying the alloying elements, using copper, zinc and magnesium. Methods of jointing were developed to augment riveting. Spot welding was used, though not in primary structures, and the main progress was in the use of rapidly applied blind fasteners and bonding.

Aircraft designers soon realised that to increase the strength of aluminium alloys without a corresponding increase in the relatively low modulus of elasticity could lead to problems in the fields of buckling, fatigue, vibration, deflection and aero-elasticity. Consequently these subjects were in the forefront of structural research between and after the two world wars. It was soon recognised that the 'allowable stress' notion of design was very unsatisfactory. What was adequate for the designers of steel and wrought iron structures in the nineteenth century was far from suitable for engineers in the aeronautical world who were trying to accommodate safety, speed, economy and efficiency into their structures in a very hostile environment. The importance of ductility and the crack-free redistribution of high stresses around rivets, for example, indicated that an ultimate and serviceability limit state philosophy would be needed if progress was to be made. Matching the ultimate resistance of components to factored loads and matching the behaviour of components to acceptable levels of deformation and vibration were the true measure of the designers' craft. These ideas were crystallised by the 1950s into the statistically based philosophy of structural safety.

As experience with the design of aluminium structures grew it was natural for the producers of the metal to look for new markets. An obvious field was the construction industry—the design and manufacture of buildings, frameworks, bridges, and smaller components such as windows, doors and canopies. Other structural areas where aluminium could be used effectively were in shipbuilding, road and rail transport, pressure vessels, and in military engineering. The quality of structural design and testing in many of these areas was relatively backward when compared with aircraft, and still is, so it became necessary to make progress carefully. Allowable stress design was adopted, but backed up by extensive research in the universities and elsewhere into problems of instability and deformation in typical construction industry components. The lateral buckling of beams was investigated, as was the compressive buckling of struts and plates, the torsional behaviour of open sections, the response to loading of flexible space frames with secondary stresses, and the behaviour of plate girders with thin webs and transverse stiffeners. Fabrication, erection and protection were also important in structures that might not be subjected to the high level of quality assurance associated with the aircraft industry.

In challenging the use of steel for general engineering structures aluminium suffered two drawbacks, the price of the material and the cost of structural assembly. Kilogram for kilogram aluminium was still relatively expensive, so the case for its use had to be carefully examined. Structures where a large reduction in dead-weight was needed, particularly in the superstructure of road and rail vehicles, military structures, and long span frameworks and bridges, were good candidates for aluminium. The costs of fabrication and assembly were influenced by the lack of information and experience in the welding of the metal, and it was therefore in this area that much research and development was carried out in the 1950s and 1960s. The successful welding of aluminium is now an accepted feature, and modern codes of practice pay much attention to it. The heat from the welding process produces a reduction in strength properties close to welds in heat-treated alloys, and the local stress-raisers in certain types of welded detail have a damaging effect on fatigue life. The way that the heat-affected zone is dealt with in the stress analysis of designs now forms an important part of design codes of practice.

The world of general engineering structures is now acknowledging the importance of limit state design, and abandoning, not without some protests, the old ways of permissible stress. This acknowledgement is not surprising. As our methods of stress analysis become more sophisticated, higher and higher local stresses are discovered in the structure, and to apply the permissible stress philosophy in these circumstances can lead to very uneconomic results. Of course, the need for economy applies to steel and concrete as well as aluminium, and the new Eurocodes in all these materials are therefore written with a limit state design philosophy.

There are many examples of the successful use of aluminium in the constructional field. A survey was carried out in 1983 of six aluminium highway bridges in the USA and one in Canada, erected in the years between 1948 and 1963. Riveting and welding were used in their construction. It was found that no painting or major maintenance had been required for the aluminium superstructures, there was no fatigue cracking in the riveted bridges and only minor cracking in the one welded structure. The lives of all the bridges were expected to be at least 50 years.

In Britain a major military bridge system, the Medium Girder Bridge (MGB), has been in continuous service for 20 years. It is an all welded, heavy duty structure, which is assembled rapidly from component parts that can be man-handled. It is manufactured from a type of 7020 alloy. Time and money were spent to develop a version of the alloy that was very resistant to stress corrosion, but the fact that bridges were deployed world-wide in a range of temperatures and conditions, with minimal structural problems, shows that the cost of the research was justified. In the military field aluminium is also used for prefabricated trackways, support boats, bridge inspection platforms, and for the structures of combat vehicles.

In addition to the more conventional structures such as masts, towers, railway carriages and road vehicle superstructures, aluminium has been used for mosque domes in Africa and the Far East. These have been constructed in the form of double-layer space frames to give architecturally interesting buildings. No doubt in the future there will be many other architectural concepts that require the use of tubular aluminium to give space and strength. There has also been an interest in the use of aluminium for the topside structures of offshore oil rigs. Additional protective structures, if required, must be added without seriously overloading the existing structure, and aluminium is an obvious candidate material.

At the time of writing much research effort is being devoted to second generation aluminium lithium alloys, the use of which can reduce airframe structural weight by 7–15% depending on the application. Advanced alloys are being developed from wrought powder metallurgy technology, which results in a rapid solidification process. This produces alloys resistant to stress corrosion cracking and exfoliation. Aramid aluminium laminates are also under development, to combine high strength sheet with the fatigue resistance of aramid fibres. The aerospace industry is a major user of high quality aluminium premium castings, particularly in the primary structure of unmanned missiles. The use of these new alloys and processes will no doubt spread to general engineering structures in time.

2 SAFETY AND SERVICEABILITY

In all modern codes of practice structural safety is established by the application of the partial safety coefficients to the loads (or 'actions') and to the strength (or 'resistance') of components of the structure. The new Eurocodes for the design and execution of buildings and civil engineering structures use a limit state design philosophy defined in Eurocode No. 1 (common unified rules for different types of construction and material).

The partial safety coefficients for actions (γ_f) depend on an accepted degree of reliability, which is recognised as a national responsibility within the European Community. The probability of severe loading actions occurring simultaneously can be found analytically, if enough statistical information exists, and this is taken into account by the introduction of a second coefficient, ψ. The design value of the action effects (when the effects are unfavourable) is then found by taking values from γ_f dependent on the type of loading and values for ψ that take account of the chances of simultaneous loading. Experts suggest a value of γ_f of 1·35 for permanent loads, such as the dead load of bridge girders, and 1·5 for variable loads such as traffic loads or wind loading. These values are similar to those proposed in the 1978 edition of the European Recommendations for Aluminium Alloy Structures produced by Committee T2 of the European Convention for Constructional Steelwork (ECCS–CECM–EKS). The loading actions on members are found by an elastic analysis of the structure, using the full cross-sectional properties of the members.

The partial safety coefficient for actions takes account of the possibility of unforeseen deviations of the actions from their representative values, of uncertainty in the calculation model for describing physical phenomena, and uncertainty in the stochastic model for deriving characteristic codes.

The partial safety coefficient for material properties (γ_m) reflects a common understanding of the characteristic values of material properties, the provision of recognised standards of workmanship and control, and resistance formulae based on minimal accepted values. The value given to γ_m accounts for the possibility of unfavourable deviations of material properties from their characteristic values, uncertainties in the relation between material properties in the structure and in test specimens, and uncertainties associated with the mechanical model for the assessment of the resistance capacity. Typical values in recent European codes of practice for aluminium are $\gamma_m = 1 \cdot 2$ and $1 \cdot 3$, on the assumption that properties of materials are represented by their characteristic values.

A further coefficient, γ_n, is often specified in codes, and this can be introduced to take account of the consequences of failure in the equation linking factored actions with factored resistance. It is often incorporated in γ_m. It recognises that there is a choice of reliability for classes of structures and events that takes account of the risk to human life, the economic loss in the event of failure, and the cost and effort required to reduce the risk.

The ultimate limit states defined by the use of the above factors refer to failure of members or connections by rupture or excessive deformation, transformation of the structure into a mechanism, failure under repeated loading (fatigue) and the loss of equilibrium of the structure as a rigid body.

Serviceability limit states, according to most definitions, correspond to a loss of utility beyond which service conditions are no longer met. They may correspond to unacceptable deformations or deflections, unacceptable vibrations, the loss of the ability to support load-retaining structures, and unacceptable cracking or corrosion. Because certain aluminium alloys in the non-heat-treated condition, or in the work-hardened condition, do not have a sharply defined 'knee' to the stress/strain curve, it is sometimes possible for unacceptable permanent deformation to occur under nominal or working loads. The same may be true for alloys that have a substantial amount of welding during fabrication.

3 HEAT-AFFECTED ZONES

In ultimate limit state design the factored characteristic loads must be shown to be less than or equal to the calculated resistance of the structure or component divided by the material factor. In calculating the resistance of welded aluminium components, however, a problem occurs with the strong heat-treated alloys. The effect of the temperature generated by the welding process is to disrupt the heat treatment and produce softened zones in the vicinity of welds. This softening is a significant factor in 6*** and 7*** series alloys, and in 5*** series alloys in a work-hardened temper. It can have a noticeable effect on the ultimate strength of the welded component and must be allowed for in design.

For many years it has been assumed that the heat-affected zone (HAZ) extends a distance of 25 mm from the centre-line of a butt weld or the root of a fillet weld. This rule originated in the USA, and was independent of the properties of the parent material, the plate thickness or the welding parameters. It was used in conjunction with the approximation that at the 25 mm line the softened zone properties jumped suddenly to full parent metal strength. It has now been shown, however, that for thicker plates and for multi-pass welds made without adequate temperature control, the zone could extend up to two or three times the 25 mm dimension. This has a noticeable effect on the ultimate strength of the member for certain weld configurations.

Research was carried out in Britain at Cambridge University, and reported in a paper to the Third International Conference on Aluminium Weldments at Munich in 1985 by J. B. Dwight and I. Robertson. They showed that the extent of the HAZ is affected by the metal temperature when welding begins, and by the build-up of temperature in multi-passes. They also found that when neighbouring parallel welds are laid simultaneously the extent of their combined HAZ increases. For thicker material the extent of the HAZ measured radially from all points along the edge of a weld was found to be proportional to $\sqrt{A_W/N}$, where A_W is the total section area of the weld deposit per pass and N is the number of heat flow paths adjacent to the weld. The extent was increased by a factor β if temperature build-up was allowed to take place between passes. This factor was shown to increase from 1 to 1·7 as the interpass temperature increased from 40°C to 150°C when welding 7*** series alloys; and from 1 to 1·4 as the temperature increased from 50°C to 150°C for 6*** series alloys.

For thinner material the extent of the HAZ measured radially from the centre-line or root of a weld was found to be proportional to A_W/N and inversely proportional to the mean thickness of the heat flow paths. The extent was increased by a factor α if temperature build-up occurred between passes, although for thin material multi-pass welding is less likely to be required. α was shown to increase from 1 to 3 as the interpass temperature increased from 40°C to 150°C for 7*** series alloys; and from 1 to 2·2 as the temperature increased from 50°C to 150°C for 6*** series alloys.

A further finding from this research was that when a weld is located too near the free edge of an outstand the dispersal of heat is less effective, and the degree of softening of the alloy is increased. The effect depends on the ratio between the distance and the weld to the free edge and the extent of the HAZ calculated as if there were no free edge effect.

The conclusions of this research have been incorporated by J. B. Dwight into the design rules for the static resistance of members in the new British code of practice, BS 8118. The HAZ is assumed to extend a distance z in any direction from a weld, where $z = \alpha \eta z_o$. The basic distance, z_0, for an isolated weld laid on unheated material with complete interpass cooling is the lesser of the two values (a) and (b) determined thus (values in mm):

(a) All types of weld	7*** series alloys:	$z_o = 30 + t_A/2$
	other alloys	$z_o = 20 + t_A/3$
(b) In-line butts only	7*** series alloys:	$z_o = 4·5 t_A$
	other alloys:	$z_o = 3 t_A$
other types of butt welds and all types of fillet weld	7*** series alloys:	$z_o = 4·5 t_B^2/t_A$
	other alloys:	$z_o = 3 t_B^2/t_A$

t_A is the lesser of $0·5(t_B + t_C)$ and $1·5 t_B$. t_B and t_C are the thickness of the thinnest and thickest elements connected respectively.

Note that for typical joints where the elements to be connected have the same thickness, $t_A = t_B = t_C$. In this case the above proposals infer that for fillet welds on material 10 mm thick, for example, the HAZ in 7*** series material is noticeably greater in width than the 25 mm often assumed in less exact codes.

The factor α allows for the possibility that at the start of deposition of a weld pass the parent metal can be at an elevated temperature due to pre-heating. The pre-heating could of course be due to the laying down of an earlier weld pass in a joint where the size of the weld requires several passes.

For a single straight continuous weld having a deposit area <50 mm^2, and t_C <25 mm, $\alpha = 1\cdot0$. If the deposit area is >50 mm^2 and t_C is >25 mm, $\alpha = 2\cdot0$. This suggests that for very large single welds on material greater than 25 mm thick, the basic HAZ width (z_o) is doubled, which illustrates the importance of proper temperature control during fabrication.

The factor η allows for an even greater heat build-up when there is a free edge, or other welding, in the same area, and where the number of paths over which heat can be dissipated is small. In the worst circumstances, η can be as high as 1·33 or 1·5.

These are fairly severe penalties, so the code allows some mitigation when the above proposals are unusually pessimistic. This involves more detailed calculation, suitable for the later, refining stages in a design. Experimental determination of the extent of the HAZ is also permitted, by conducting hardness surveys.

In spite of the findings of this research, and the incorporation of the findings in BS 8118, as indicated above, there is still considerable resistance in the international structural aluminium community to a departure from the 25 mm rule. There is a school of thought that by the time all mitigating circumstances are taken into account, and when the heat-affected material only occupies a relatively small area of the structure, the inaccuracy of the 25 mm rule has only a very minor effect on the calculation of total structural resistance. The policy of BS 8118 is to present the designer with all the facts, and the possibility of using more simple, but perhaps pessimistic rules during the initial design stage. At later stages more refined, less pessimistic recommendations can be pursued.

4 TENSION MEMBERS

The static design problems of axially loaded tenison members are not considered to be of great complexity, and research in this field has been very limited in recent years. In recent codes of practice the main task has been to transfer the design method from 'allowable stress' to limit state, without introducing any new fundamental analysis of behaviour. Thus the requirement is now that the direct tension in a member under factored loading does not exceed its factored tension resistance.

For the simplest tension member, with no welding or holes to affect the strength of a cross-section, the factored resistance is given very simply by the expression pA/γ_m, where γ_m is the partial safety coefficient for material properties, A is the area of the cross-section, and p is the limiting stress in tension.

The limiting stress in tension depends on the relationship between $f_{0\cdot2}$, the guaranteed minimum tensile 0·2% proof stress of the material, and f_u, the ultimate tensile strength of the material. The UK code, BS 8118, proposes that, when $f_u < 1\cdot4\, f_{0\cdot2}$, $p = f_{0\cdot2}$; and when $f_u > 1\cdot4\, f_{0\cdot2}$, $p = 1\cdot28\, f_{0\cdot2} - 0\cdot2 f_u$. This selectivity is to guard against the possibility that for materials with a very marked elasto-plastic behaviour between proof and ultimate conditions, there could be some permanent elongation of members under working stresses.

If the section of a tension member contains HAZ material, there will be a reduction in strength due to the reduced properties of the material in the zone. In the simplest terms, if the area of the zone is A_z, then the factored resistance is now $(p(A - A_z) + p_z A_z)/\gamma_m$, where p_z is the limiting stress in tension of the heat affected zone material. An alternative, used in BS 8118, is to designate p_z/p as k_z and rewrite the factored resistance as $p[A - A_z(1 - k_z)]/\gamma_m$. The expression in the brackets is then the effective section area, called A_e. A further possibility, also used in BS 8118, for members having a uniform thickness material in their cross-section, is to consider the HAZ zone as having an effective thickness t_z, where $t_z = k_z t$. It is a matter of opinion as to how far these 'simplifying' procedures should be taken. The danger in dimensional simplifications is that they often only apply in certain circumstances, whereas the basic concept of a reduced strength of material in the HAZ is always correct and clear.

5 COMPRESSION MEMBERS

Much relatively recent research in this area has been concerned with the effects of high residual stress, and of welding, on the form of the classic strut curves linking the limiting stress of overall buckling with slenderness ratio. A second area of interest has been the buckling

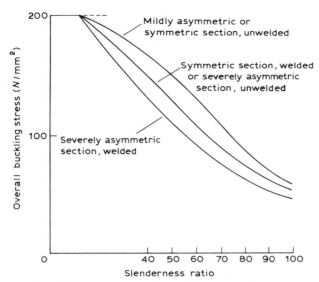

Fig. 1. Examples of strut curves from BS 8118, drawn for a section having a compressive strength of 200 N/mm² at zero slenderness ratio (Dwight). Note: A mildly assymetric section has $Y_1/Y_2 \leqslant 1\cdot 5$, where Y_1 and Y_2 are the distances from the buckling axis to the further and nearer extreme fibres, respectively.

stress for thin plates, and the effect of this on the local buckling of thin-walled compression members. The proposals of J. B. Dwight, adopted in BS 8118, are that for most design problems three curves are sufficient for overall buckling, covering two categories of section geometry, 'symmetric or mildly asymmetric' and 'severely asymmetric', and two categories of construction, unwelded and welded. The strut curves in these proposals, examples of which are shown in Fig. 1, are based on the 'plateau' concept of a constant limiting stress at very low slenderness ratios, followed by a diminishing limiting stress as the slenderness increases in the form of a Perry curve. The categories of geometry and construction are catered for by varying the constants in the Perry formulation.

Because of the effects of initial deviation from straightness, variable material strength and other unknowns, care has to be taken that the design curves do not imply a level of scientific accuracy that does not exist, but it is generally thought that three separate curves embracing the categories discussed above is a reasonable compromise between theory and practice. The value of limiting stress at the plateau (p_1) is governed by the geometry of the cross-section of the strut. If there is no premature local buckling of the elements then the full 'squash' load will be realised and the section is designated 'compact'. If local buckling lowers the resistance of the strut because of premature instability of the thin elements, the section is designated 'slender' and the value of the limiting stress at the plateau is reduced.

There are a number of ways of calculating the limiting local buckling stress for slender sections. An exact analysis of the section, taking account of full interaction between the component flat or curved elements, can be made, but this is time consuming and demanding. It is more usual to divide the cross-section into internal or outstand elements either unreinforced or reinforced by the provision of stiffening ribs or lips. Internal elements are fully supported along longitudinal edges, outstand elements are supported along one longitudinal edge, and free (or supported by a reinforcing lip) along the other. The classification of the complete section in local buckling is then taken as that for the least favourable element, and this approximate procedure greatly simplifies the analysis. The limiting local buckling stress of this element is taken as that for the entire cross-section.

In order to make the procedure more automatic, simpler to apply, and more easily set down in design code form, the effect of local buckling in slender sections can be thought of as equivalent to a reduction in the cross-sectional area of the critical element. For elements of uniform thickness it can be thought of as a reduction in thickness. This has no physical meaning, and used unwisely can lead to confusion and subsequent error. However, for better or for worse the reduced thickness concept has been used in BS 8118, where the limiting stress at the plateau for slender sections is taken as $p_1 = (A_e/A)p_o$. The value of p_o is that for overall yielding of the material, and A_e is the area of the effective section taking account of conceptual area or thickness reductions to allow for local instability. Thickness reduction is governed by the coefficient k_L, where effective thickness $t_e = k_L t$. The coefficient k_L is in general a function of β/ε, where

$$\varepsilon = \sqrt{250/p_o}$$

and β is a function of b/t, where b and t are the width and thickness respectively of the least favourable thin-walled element.

Typical relationships between k_L and β/ε are shown in Fig. 2 for

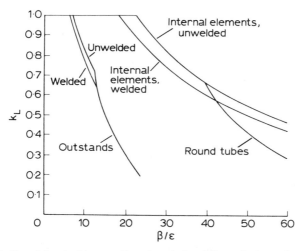

Fig. 2. Relationship between the local buckling factor, k_L, and the parameter β/ε (Dwight) in BS 8118.

welded and moulded internal and outstand elements, and for round tubes.

So far, our consideration of local buckling in slender cross-sections has not included the effect of welding. If an element is both slender and affected by HAZ softening, the reduced thickness in the softened area must be taken as the lesser of $k_L t$ and $k_Z t$, where k_Z was defined in Section 4 above.

6 FATIGUE

The fatigue of aluminium structures first became important in the aircraft industry, where the need for low weight meant that designers had to design components to operate at stresses where fatigue failure would limit the life of the structure. This was a major departure from heavier construction in steel, for example, where allowable stresses often fell below the level of the fatigue limit of the material. A further problem was that the curves limiting cyclic stress to the number of loading cycles for aluminium structures continued to show a slight negative slope at long lives.

Research in Europe in the nineteenth century had shown that the effect of mean stress on life, in addition to alternating stress, could not be ignored. Constant life contours were drawn showing the interaction between alternating and mean stress by Goodman in his book on *Mechanics Applied to Engineering,* in 1899, and this method of summarising test information was widely used by aircraft designers. Standard shapes were suggested for the contours, but these did not take sufficient account of the statistical nature of most fatigue test information.

Much of the early test work was on highly polished specimens in a laboratory environment, but it was soon realised that corrosion could affect fatigue performance. Furthermore, in the aircraft industry most connections between components used riveted or bolted construction which introduced local stress concentrations and involved sudden changes in shape. Thus the influence of stress concentrations on fatigue became a major area for research, since fatigue failures usually originate at discontinuities. The problem of assessing the fatigue life of components that have a number of detailed features, each with its own stress concentration level, became important. The complete component was found to have an $S-N$ curve that represented the lower envelope of the $S-N$ curves of the detailed features, and this highlighted the problem of predicting the failure of components or complete structures from the results of laboratory tests on small specimens containing only the stress-raising feature.

The other problem tackled by the aircraft industry was the effect of a spectrum of variable amplitude loads on fatigue life, and the concept of failure due to cumulative damage. The work of Palmgren in 1924, and by Miner about 20 years later, is well known.

When the aluminium industry began to explore the use of high strength alloys in general engineering structures, the whole of the extensive research data on fatigue in aircraft structures was available. There was a difficulty, however, in that little of this data could be applied to welded structures. The stress concentration effects of typical welded details were not easily assessed, and the welding process introduced residual stresses in structural components. Design information, therefore, tended to follow the lead given by the steel codes of practice, in that fatigue stress levels were related to the particular type of welded detail. The details were classified and curves were presented that showed the relationship between stress range, life and mean stress for each classified detail. Information of this type was given, for example, in the British Code of Practice CP 118 (1969), where 18 graphs and 9 tables were required.

More recently there have been changes to the way that design information has been presented, with particular reference to the influence of residual stresses due to welding and to the importance of weld defects. In welded details of both steel and aluminium the crack initiation phase is insignificant and the fatigue life is mainly governed by the propagation of a crack from the weld root or weld toe. For a given geometry, however, the applied stress to produce a known growth rate in aluminium is one third that of steel. This leads to the proposal that for a given joint detail the fatigue strength in aluminium is one-third that in steel. A further simplifying proposal, suggested by Maddox of the British Welding Research Institute, results from the fact that welded joints in aluminium contain very high residual stresses. The effect of these, when combined with applied stress, is to make fatigue strength virtually independent of the mean stress. It is suggested that design rules, like those for steel structures, need be based only on stress range, since critical areas, such as weld toes, would be cycling from yield tension downwards.

How far these simplifications can be supported by test results is a matter of debate. For certain details there seems to be good agreement, but in others there is still doubt about the accuracy of the idea. There are so many unknowns in the collation of test results—the condition of the specimens, the standard of the welding—that it is difficult to reach sound conclusions. There are also difficulties with the cumulative damage predictions when one of the variable amplitude loads occurs at the low stress–high cycle end of the spectrum, because the shape of the $S-N$ curve at cycles beyond 10^8 is rarely known. It seems likely that designers will have to take a fracture mechanics approach to get sensible results. Cullimore, in a recent study, has shown that there can be some discrepancy between the predictions of the British Codes of Practice, CP 118 and BS 8118. The Canadian Standards Association code 'Strength Design in Aluminium' suggests that welded aluminium details have a fatigue limit beyond 5×10^6 cycles for constant amplitude loading. A recommended specification submitted to the Aluminium Association of the USA makes a similar proposal, but beyond 10^7 cycles.

The assessment of reliability also varies with the codes of practice. Kosteas and Poalas have proposed an expression of the type $\Delta\sigma_R/\gamma_m \geqslant \Delta\sigma_e \cdot \gamma_s$, where $\Delta\sigma_R$ is a characteristic value of the fatigue strength distribution, $\Delta\sigma_e$ is the mean value of the loading stress distribution, and γ_m and γ_s are partial safety factors. The British code BS 8118 uses the notion of a fatigue life factor γ_L, and a fatigue 'ease of inspection' factor (γ_{mf}). The former can be applied at the discretion of the designer to the nominal design life, the latter to the allowable stress range, and this design philosophy is particularly applicable to damage tolerant structures, where stable crack growth between inspections is allowed.

The choice of γ_L could be influenced by the possible effects of increasing crack growth during the later stages of the life of the detail, the accuracy of the assumed loading spectrum, whether records of loading are kept, and the chance of a change of use of the structure in mid-life. A designer, wishing for maximum economy at the design stage, might wish to ignore all these possibilities, and set γ_L at 1·0. If he wishes to avoid trouble in the later stages of the fatigue life (which might be 70 years away), he might make $\gamma_L = 1\cdot 25$. If he knows nothing about the future use of the structure he could play very safe and make $\gamma_L = 5$. The code does not make any recommendations for the value of γ_L, which is considered to be a matter for the designer.

The choice of γ_{mf}, known in BS 8118 as the fatigue material factor, could be influenced by the need for the detail to exist in a very hostile environment, and whether, in the event of detail failure, alternative load paths exist. It could also be influenced by the ease with which the detail can be inspected for cracks. As in the case of γ_L, the code does not make recommendations for γ_{mf}, but a designer wishing to take account of these factors could apply a value of 1·2, for example.

The relationship between stress range, f_r, and endurance, N, for design purposes has been set down in BS 8118 by M. Ogle. It consists of a family of curves, the generalised form of which is shown in Fig. 3, plotted on logarithmic scales. The design curve represents the mean minus 2 standard deviation level below the mean line through experimental data. The constant amplitude cut-off stress, f_{oc}, occurs at 10^7 cycles, and at stresses below this constant amplitude stress cycles are assumed to be non-damaging. The variable amplitude cut-off stress, f_{ov}, occurs at 10^8 cycles. The slopes of the f_r–N curves are changed from $1/m$ to $1/(m+2)$ between 5×10^6 and 10^8 cycles. Typical welded and unwelded details are classified, by giving them a class number which specifies a reference strength, f_r, at 2×10^6 cycles. The choice of class depends upon many factors including product form, type of weld, the way the weld is laid, and the ratio of

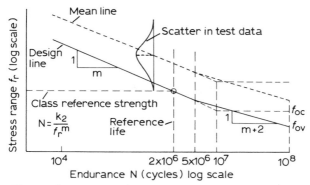

Fig. 3. Generalised fatigue curves (BS 8118) (Ogle).

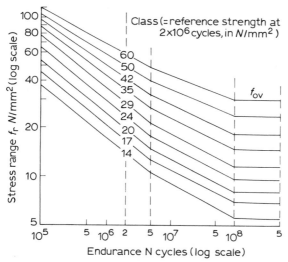

Fig. 4. Design f_r-N curves for variable amplitude stress histories, BS 8118 (Ogle).

thicknesses of the elements to be welded together. The proposed design curves for BS 8118, reproduced with permission of the British Standrds Institution, are shown in Fig. 4.

7 FRACTURE MECHANICS (M. S. G. Cullimore)

From the earliest days of calculated design the engineer assumed that the components of his structure were to be made of homogeneous and isotropic materials, and perfectly constructed. Knowing this not to be so he used a large factor of safety to limit the risk of failure. Experience of structures in service, refinement of analyses, testing and improvements in materials quality has gradually narrowed the gap between the idealised and the actual so that, with discernment and judgement, safety factors have been derived which are adequate but not excessive.

Experience has shown that real structures, however good the material and however carefully fabricated, will almost inevitably contain cracks, or defects which are sufficiently sharp to be treated as cracks. The study of the behaviour of material containing cracks, fracture mechanics, seeks to explain the role of such defects in reducing the strength of components below that based on general yielding of the assumed perfect component and to provide quantitative answers in specific cases. For most practical purposes the effect of the crack can be separated into a geometrical effect and a toughness effect: the first is common for all materials and the second is a measurable property for the particular material.

Although the stress at the tip of a sharp crack is undeterminable by classical stress analysis methods, its severity may be described in terms of a 'crack tip stress intensity factor' (K). For a flat plate containing a through-thickness crack of length $2a$ (Fig. 5) which is subjected to an average tensile stress σ, K is given by the expression

$$K = \sigma\sqrt{\pi a} \cdot f(a/w) \tag{1}$$

where $f(a/w)$ is a dimensionless parameter depending on the geometries of the component and the crack, commonly denoted by α. As the width of the plate increases $f(a/w) = \alpha \rightarrow 1 \cdot 0$.

Failure occurs when the stress intensity $\sigma\sqrt{\pi a}$ reaches a 'constant' critical value, K_c, which may be measured experimentally. Values of $f(a/w)$ may be calculated analytically for other component geometries

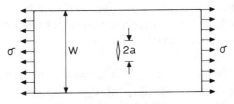

Fig. 5. Centre cracked flat plate.

and the value of K_c measured for the material can then be used to predict the critical combination of stress and crack length. The stress intensity factor characterises the crack and so data obtained from a simple test specimen may be applied to predict the behaviour of components of the same material having the same value of K.

The derivations on which the theory of 'linear-elastic fracture mechanics' (LEFM) is based assume linear-elastic behaviour of the material. For the fracture of high strength metallic materials under plane strain, where the general level of stress is below the elastic limit, although the very high stress concentration there will produce a zone of plasticity around the tip of the crack it will be of negligible size, so that general linear-elastic conditions may be assumed, and LEFM methods used. For plane stress conditions, or with more ductile materials, the size of the plastic zone may not be negligible, effectively increasing the severity of the crack. This effect may be allowed for by applying an empirical correction which increases the length of the crack so that the LEFM solution gives approximately the correct result.

The critical stress intensity factor for fracture (K_c) will depend on the constraint of strain in the material in the crack tip region which is related to the thickness of the component, as indicated in Fig. 6. It is largest for plane stress conditions where the constraint is low. Constraint increases with increasing thickness, as plane strain conditions are approached, and K_c tends to a constant limit value, K_{Ic}, known as the Plane Strain Fracture Toughness, which can be treated as a material property.

Internationally recognised standard tests have been devised for measuring K_{Ic} using notched specimens loaded either in tension or in

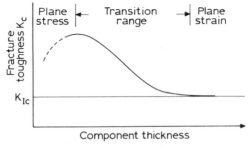

Fig. 6. Relation between fracture toughness and component thickness.

Table 1. Values of K_{Ic} for Materials in Plate Form

Material	$f_{0.2}$ (MN/m^2)	K_{Ic} (MN/m$^{3/2}$)
Aluminium alloy 2124-T851 (Cu–Mg)	443	35·8
Aluminium alloy 7071-T73651 (Zn–Mg)	444	42
Low alloy steel BS EN24 (Ni–Cr)	1180	95
Low alloy steel ASTM A533B (Mn–Cr)	480	153

bending. To obtain conditions of plane strain, so that a valid K_{Ic} value is obtained, it is generally accepted that the thickness of the specimen should be greater than $2·5\ (K_{Ic}/f_{0.2})^2$. Table 1 gives values of K_{Ic} for some materials in plate form, for specimens taken in the direction of the grain flow and notched transversely to it.

The values of K_{Ic} in Table 1 were obtained from tests done at room temperature (10–22°C). Variations in temperature which affect the proof stress of the material also affect K_{Ic}. For example the proof stress of the Mn–Cr steel in Table 1 had increased to 600 MN/m^2 at -100°C whilst the value of K_{Ic} had dropped to 44 MN/m$^{3/2}$. Fracture toughness is also affected by strain rate, reducing as the strain rate increases.

In fatigue loading where the stresses generally are low compared with the yield stress of the material, and where the crack tip plastic zone is comparatively small, LEFM methods may be used to describe the crack behaviour.

For constant amplitude fatigue loading it is found experimentally that the crack growth rate (da/dN) for a particular material is the same function of stress intensity factor range (ΔK) and stress ratio $(R = f_{min}/f_{max})$ for different stress ranges and for different geometries.

Crack growth data comprise measurements of crack length (a), made at increments of N (cycles) during a cyclic loading test on a standard test piece. The average rate of growth of the crack $(\delta a/\delta N)$ over these intervals of N, together with ΔK for each corresponding crack length, is then calculated. A continuous crack growth curve is then constructed in terms of a logarithmic plot of (da/dN) and ΔK through the plotted points. When studying such a curve it should be remembered that, when a number of plotted points appear, they may be the results for only one specimen. The results of a number of tests may be presented as a mean curve through the plotted points for the individual tests or sometimes as a band enclosing, say, 95% of the data.

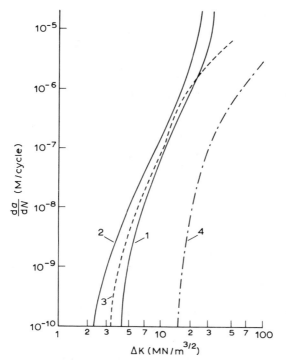

Fig. 7. Fatigue crack propagation rate curves in laboratory air. Curve 1, aluminium alloy (Cu-Mg), 2024-T3 plate ($0.05 < R < 0.10$); curve 2, aluminium alloy (Cu-Mg), 2024-T3 plate ($R = 0.5$); curve 3, aluminium alloy (Zn-Mg), 7075-T7351 plate ($0.08 < R < 0.11$); curve 4, corrosion resistant steel, ASTM-A286 forging ($0 < R < 0.05$).

Figure 7 gives some fatigue crack propagation rate data for two aluminium alloys and for a corrosion resistant steel. (These data, and those in Table 1, have been taken from Data Items 81031, 84003 and 83023, published by ESDU International plc, London.) Comparison of curves 1 and 2 shows that an increase in stress ratio (R) causes an increase of crack growth rate. Corrosive environments also increase crack growth rates. When the corrosion effect is significant the rate generally increases with decrease in load cycling frequency. This corrosion fatigue behaviour may, however, be modified if the material is susceptible to stress corrosion cracking.

Experimental crack propagation rate curves are of a generally sigmoidal form with an approximately linear part in the middle of the range, for small steps in which

i.e.
$$\log(\mathrm{d}a/\mathrm{d}N) = C + m \log \Delta K \qquad (2)$$
$$(\mathrm{d}a/\mathrm{d}N) = C(\Delta K)^m$$

where C and m are material parameters. Equation (2) is known as the Paris law.

The crack growth life of a component, that is the number of cycles required to extend a crack from a known (or assumed) initial length a_o to a larger specified length a_1, may be estimated by integrating eqn (2)

$$\frac{\mathrm{d}a}{\mathrm{d}N} = C \Delta K^m$$

over a series of discrete intervals in the range a_o to a_1. This may be done analytically in a few simple cases where m is an integer and the geometrical factor $\alpha = f(a/w)$ in eqn (1) may be expressed explicitly in terms of a; otherwise numerical methods of integration are required.

Linear-elastic fracture mechanics is adequate for discussing most cases of brittle fracture, stress corrosion cracking and fatigue crack growth, where the stress levels away from the crack are generally low compared with the yield stress. Where the crack tip plastic zone becomes large compared with the size of the crack, in ductile tearing, creep crack growth, or other cases where the overall behaviour of the component ceases to be linear-elastic the problem must be treated elasto-plastically, using the techniques of elasto-plastic fracture mechanics (EPFM) or general yielding fracture mechanics (GYFM). These techniques are not yet so widely practised as those of LEFM and some design methods, although theoretically based, involve empiricisms and lead to less exact solutions.

The rate of increase of strain in an element of a structure will depend on the type of loading. Values of the rate of increase of strain/second ($\dot{e}$) for some typical loadings are:

	$\dot{e}$
Traffic on bridges	10^{-5} to 10^{-3}
Wave slamming on ships' plating and Snap-over buckling	2×10^{-3} to 6×10^{-1}
Collision—ships' plating (local)	2
Shock loading	>20

When $\dot{e} < 10^{-3}$ stresses can generally be calculated assuming the load to be statically applied; for higher strain rates the dynamic response of the structure to the loading must be taken into account in the calculations.

The strengths and fracture properties of the material of the element may also be changed by the higher strain rates. For example, the critical dynamic fracture toughness of a structural mild steel was found to be halved when the strain rate increased from 0·8/s to 3·8 × 10⁴/s. However, the toughness of aluminium alloys, which do not fracture by cleavage, is relatively insensitive to strain rate. This aspect of the fracture behaviour of aluminium alloys appears to have attracted little attention, and data on high strain rate fracture properties are very limited.

The fracture mechanics of fast fracture and crack arrest is a highly specialised topic and the determination of dynamic stress intensity factors is a very difficult problem, involving a combination of analysis and very sophisticated experimental techniques.

Using fracture mechanics it is possible to calculate the residual strength of a component containing a crack of known size or, alternatively, the largest crack that can safely be tolerated under service loading: under fluctuating loading the time taken for an existing crack to grow to a specified limiting size may be estimated. Thus the residual strength of damaged structures may be assessed, material and fabrication tolerances specified and crack monitoring schedules set for damage tolerant designs. It follows that non-destructive testing (NDT) and fracture mechanics are essentially complementary and that continuing improvements in crack detection and monitoring techniques will increase the reliability of fracture mechanics estimations. Fracture mechanics is now established and extensively used, both in design and in service, in the aerospace, gas, petroleum, marine, chemical, nuclear and offshore industries. In all but the first of these problems arise mainly from welding defects.

The general practitioner of structural engineering will rarely use fracture mechanics directly in design. The rules in modern codes of practice—used for example in specifying material toughness for avoidance of brittle fracture and design against fatigue—will, however, have been devised using fracture mechanics principles.

Finally it would be wrong to conclude this section without mentioning probabilistic fracture mechanics. The impetus for this relatively new branch of fracture arose in the nuclear industry from a demand for probability estimates of pressure vessel integrity. The detailed assessment considers how the failure probability of a vessel is related to the statistical distribution characterising the incidence and size of defects in weld seams.

The statistical principles used are not new; the difficult part of the problem is knowing how to treat the statistical distribution of the incidence and size of defects. Models have been developed which lead to the expression for the failure probability of a structure (made of a brittle material in which the stress field is non-uniform) and account for the presence of many different types of defect. They can be used to investigate the effects of fracture toughness variations and various defect geometries, and can account for the growth and nucleation of defects in service resulting from fatigue, stress corrosion or creep.

8 WELDED JOINTS

The principles of welded joint design look to be coming close to those for welded steel joints. The main difference is that the strength of the heat-affected zone material in the adjacent member has to be checked, as well as the strength of the weld metal itself. Some earlier codes of practice gave allowable design stresses for welded joints but did not make clear whether failure might be expected in the weld itself or in the adjacent material of the structural member. Part of the trouble was the lack of clarity in defining a joint. It is now recognised that a joint, or connection, which can be rivetted, welded, bolted or bonded, can join the basic elements of a component together to form the cross-section (e.g. the web to the flange of a plate girder); or it can join complete structural members together to form a framework (e.g. beams to columns in a building frame); or it can join local brackets to members; or join eyes and jaws to the ends of members to enable rapid assembly in the field to be achieved.

In all these cases a proper stress analysis has to be made of the jointing medium. When this is a weld great care has to be taken to ensure that the leg length, throat thickness and weld quality are adequate to provide the strength of the connection. The analysis given in BS 8118 has been checked experimentally, and the details are described by Soetens in a subsequent chapter. The experimental studies are on carefully manufactured details in a laboratory environment, so some feature needs to be introduced into the analysis to account for welds that will be in a hostile environment during their life, or where adequate quality control cannot be ensured. A proposal is that these circumstances can be taken into account by increasing the value of the partial safety factor for the material (γ_m).

Looking back over the past 20 or 30 years it is noticeable that most of the research expenditure has been devoted to the analysis and testing of members in bending, compression and tension. Joint design, however, which is less amenable to elegant mathematical analysis, has been neglected. The work reported by Soetens is about the only fundamental contribution available from Europe. A further area of neglect is the relationship between the joint design rules and the tests to confirm the competence of welders. Very often the welder has to produce a simple joint that meets a strength requirement, but the rules do not distinguish between failure of the specimen in the weld or in the adjacent heat-affected material.

9 TESTING AND QUALITY ASSURANCE

It is assumed that sophisticated testing facilities and quality assurance techniques of the type used in the aircraft or aerospace industries are too expensive for the average deisgner or client in general engineering structures. It is necessary, however, for the designer to be aware of testing techniques and procedures which may be used in place of analysis and calculation to prove the resistance of a structural member. It is also important that the designer can judge the quality of a test, and be aware of the limitations of test equipment. For some reason the validity of test data is often unquestioned, whereas endless discussions take place on the accuracy of analysis.

Verification of structural resistance by test is sometimes necessary if there is doubt about the design procedure, or the material characteristics, or the quality of fabrication. In certain instances there are no adequate analytical methods, and verification by test is the only way of proceeding. Ideally, enough tests should be carried out for a statistical analysis to be made of the mean resistance and the standard deviation of experimental scatter, but for many structures and components this would be too time consuming and expensive. Considerable engineering judgement is required by the designer in assessing the cost-effectiveness of a testing programme.

It is not always easy to reproduce in a test facility the precise loading conditions or conditions of support and restraint that the structure will experience in practice. In many cases the test will be carried out under the conditions assumed in the design calculations, and the test then checks the accuracy of the analysis based on these assumptions. An alternative 'acceptance test' approach checks that the structure will adequately carry out the task for which it is required, irrespective of the conditions assumed for design purposes. The cost of 'acceptance' testing, however, can be very high. In the aeronautical field the cost of testing is now so high that there are pressures to certify the structural adequacy of aircraft by calculation alone; particularly if the structure to be tested is a modification of a previously tested structure.

It must be remembered that in non-linear structures the order in which loads are applied can influence ultimate resistance, and that in certain types of structure the deflections and deformations of the members can affect the direction and magnitude of the applied loads. Under combined loading the proving test would normally be arranged so that the lowest resistance is measured. The principle of operation of the test rig is also an important factor. The gradual increase of static loading by suspending dead weights results in progressive failure until at the failure load the structure deforms without reduction in applied load. The increase in loading by jacking the structure against an external framework results in loss of stiffness as failure is initiated, with a consequent falling-off in load. Loads applied in this way mean that little energy is stored, and this form of loading is very common. Care is required, however, to ensure that the methods of attaching screw jacks or hydraulic jacks to the structure do not locally increase structural strength or stiffness, or result in local failure.

The procedure for static load application can vary, but methods suitable for aluminium structures are given in BS 8118. It is proposed here that before applying the desired load combination the structure should be loaded and unloaded once to settle the structure, and that the load level for this should not exceed the unfactored loading or a load that would cause the deflection serviceability limit to be reached. This load is to be maintained for 15 min, and after removal a further period of 15 min should elapse before proceeding with the main test. Unfactored design loads should be applied in five equal increments. The fifth increment should be maintained for 15 min, since this is the serviceability limit load. All loads are then removed and the structure inspected. The structure is then loaded incrementally to the full factored design load, otherwise known as the 'proof load', recording deflections and strains. The increase from unfactored to factored load should be made with at least five increments, and the full factored load should be maintained on the structure for 15 min. After removal of this load the residual deflection of the test specimen should be recorded. This deflection should not exceed 5% of the deflection under full load. If it is in excess of this figure the test may be repeated. If, after a further ten applications of the proof load this requirement is still not met, the structure is deemed to have failed the test.

For type testing a product an Ultimate Resistance test is often specified. This is usually done by applying the proof load in one increment, and then gradually increasing all imposed loads proportionally until failure occurs.

Structures or components with high stress raisers are susceptible to fatigue, as discussed earlier, so there is often a good case for a fatigue test. It may be necessary to obtain a stress range versus cycles to failure relationship for a structural detail at the design stage, because the detail does not match any of the details for which fatigue information is given in the design code. If this is so, then BS 8118 proposes that a minimum of eight identical specimens should be tested to give endurances in the range 10^5 to 10^7 cycles. A mean curve should be calculated from the results, and a design curve established which is parallel to the mean curve and two standard deviations away (or 80% of the mean curve, whichever is the lower). This allows for the possibility of wide variations in fatigue strength in production—wider than would normally be expected from eight specimens.

The criteria for acceptance depends on whether the structure is required to give a 'safe-life' or 'damage tolerant' performance. In safe-life design the life to failure (adjusted to take account of the number of test results available) should not be less than the factored design life, i.e. $\gamma_L \times$ nominal design life.

The quality assurance of welded structures is important, because poor welding, out-of-tolerance dimensions, scored and dented members and incorrect fabrication techniques can lead to premature failure either statically or under cyclic stresses. There have been a number of failures of aluminium lifting frames that have been attributed to poor workmanship. It is therefore essential to establish proper welding procedures, and to test welders to these procedures. Inspection of finished structures should be carried out using radiographic or other non-destructive techniques, and use can be made of dye penetrants.

Welding should normally be carried out using the inert gas shielded metal arc (MIG) process or the inert gas shielded tungsten-arc (TIG) method. If the latter is used special care must be taken to limit the temperature generated. If this is not done the TIG process will result in a reduction in weld strength, and this possibility must be foreseen during the design stage. The British code design rules are given for the MIG process only, but the codes of mainland Europe often give ranges of weld strengths for both processes.

10 CONCLUSIONS

There seems to be no reason why the principles of design discussed in this chapter cannot be incorporated into a Eurocode for Structural Aluminium. These are various interpretations of safety and reliability in Europe, for example, but the fundamental process is common. The analysis of structural members to establish their ultimate static resistance is similar everywhere, due to the wide and early distribution of the findings of research programmes through journals and conferences. There may be local variations but these do not lead to significantly different structural economies. The comprehensive strut curves recommended by West German and Italian codes of practice are slightly different from those given in British codes, but the differences are not vital and it would be straightforward to come to a common European format. Beam and plate girder design has already been unified for steel construction in Eurocode No. 3, and much of that philosophy is applicable to aluminium, providing the heat-affected zone effects are properly taken into account. Considerable efforts are being made to approach a single method of assessing the effects of fatigue.

In parallel with the work towards a Eurocode the International Standards Organisation is also drafting a code of practice for the design of general aluminium structures. This brings countries such as the USA, Canada and Sweden into the field. It would be a considerable advantage if the ISO and European philosophies could be harmonised and we must hope that this becomes a possibility during the next few years. What is reassuring is that the experts from most of the aluminium-using countries are in closer contact now than for a long time, and a sense of constructive co-operation is in the air.

In assembling this review of design principles the writers became aware of the need for a bibliography covering the many fields where aluminium structures are in use. The chapter therefore concludes with an attempt to bring together some of the key papers, reports and books on the subject. The references to fracture mechanics apply to both steel and aluminium.

BIBLIOGRAPHY

Armstrong, R. A., Design and fatigue life prediction technique for aluminium joints. 3rd International Conference on Aluminium Weldments, Munich, April 1985. Aluminium-Verlag, Dusseldorf.

Baehre, R. & Riman, R., Traglastuntersuch ungen von Druckstäben aus Aluminium mit Quernähten. *Schweissen und Schneiden,* **38,** (1986).

Baker, J. F. & Roderick, J. W., The strength of light alloy struts. Aluminium Development Association Research Report No. 3, 1948.

British Standards Institution, British Standard Code of Practice CP 118: 1969, The structural use of aluminium. London, 1969.

British Standards Institution, British Standards BS 8118, Code of practice for the Structural Use of Aluminium. London, 1992.

Brungraber, R. J. & Clark, J. W., The strength of welded aluminium columns. *Proc. ASCE,* Journal of Structural Division (August 1960).

Bulson, P. S., PhD thesis, University of Bristol, 1953.

Bulson, P. S., Local instability problems of light alloy struts. Aluminium Federation Research Report 29, December 1955.

Bulson, P. S. & Nethercot, D. A., The new British code for the design of aluminium structures. Coll. on thin-walled metal structures in building, IABSE, Stockholm, 1986, p. 43.

Canadian Standards Association, Strength design in aluminium, CAN3-S157-M83. Rexdale, Ontario, 1983.

Cappelli, M., de Martino, A. & Mazzolani, F. M., Ultimate bending moment evaluation for aluminium alloy members: A comparison among different definitions. *Proc. Int. Conf. on Steel and Aluminium Structures,* Cardiff, July 1987. Elsevier Applied Science, London, p. 126.

Chilver, A. H. & Britvec, S. J., The plastic buckling of aluminium structures. *Proc. Symp. Aluminium in Structural Engineering.* Institution of Structural Engineers, London, 1963.

Clark, J. W., Eccentrically loaded aluminium columns. *Trans. ASCE,* **120,** (1955) 116.

Clarke, J. D. & Swan, J. W., Interframe buckling of aluminium alloy stiffened plating. ARE Report AMTE(S) R85104, October 1985.

Clarke, J. D., Buckling of aluminium alloy stiffened plate ship structure. *Proc. Int. Conf. on Steel and Aluminium Structures,* Cardiff, July 1987. Elsevier Applied Science, London.

Cullimore, M. S. G., Aluminium double angle struts. *Proc. Symp. Aluminium in Structural Engineering.* Institution of Structural Engineers, London, 1963.

Cullimore, M. S. G. & Millward, C. P., Development of a fatigue resistant aluminium/steel HSFG bolted joint. *Proc. Int. Conf. on Steel and Aluminium Structures*, Cardiff, July 1987. Elsevier Applied Science, London, p. 162.

Dier, A. F. & Dowling, P. J., Aluminium plated structures—Final Report for AMTE. Imperial College Department of Civil Engineering Report CESLIC-AP52, July 1981.

Dier, A. F., Comparison of steel and aluminium plate strengths. *Proc. Int. Conf. on Steel and Aluminium Structures*, Cardiff, July 1987. Elsevier Applied Science, London, p. 193.

Dwight, J. B., Aluminium strut design. *The Structural Engineer* (February 1961) 47.

Dwight, J. B., Aluminium sections with lipped flanges and their resistance to local buckling. *Proc. Symp. Aluminium Structural Engineering*. Institution of Structural Engineers, London, 1963.

Dwight, J. B., Use of Perry formula to represent the new European strut curves. Int. Coll. on column strength, Paris, November 1972.

Dwight, J. B., Use of the Perry formula to represent the European column curves. *Steel Construction*, **9** (1) (1975).

European Convention for Constructional Steelwork, European recommendations for aluminium alloy structures, first draft, 1978.

Evans, H. R. & Burt, C. A., Ultimate load determination for welded aluminium plate girders. *Proc. Int. Conf. on Steel and Aluminium Structures*, Cardiff, July 1987. Elsevier Applied Science, London.

Faella, C., Comportamento flessionale delle barve in alluminio al di la dei limiti elastici. *La Ricerca*, 1976.

Fracture mechanics in design and service 'Living with defects'. *Phil. Trans. R. Soc. London, A*, **299** (1981) 1–239.

Cox, H. L. *et al.*, (eds). *A General Introduction to Fracture Mechanics*. Mechanical Engineering Publications, London and New York, 1978.

Gilsons, S. & Cescotto, S., Experimental research on the buckling of aluminium alloy columns with unsymmetrical cross section. ECCS-TC 2 Report, 1981.

Gunn, K. W. & McLester, R., Fatigue strength of welded joints in aluminium alloys. *British Welding Journal*, **9** (12) (December 1962).

Harrison, J. D., Gurney, T. R. & Smith, B., Cumulative damage of fillet welded joints in AlZnMg alloys. *Proc. Int. Conf. on Fatigue of Welded Structures*. The Welding Institute, 1971.

Hartmann, E. C., Structural applications of aluminium alloys. *Trans. ASCE*, **102** (1937).

Hartmann, E. C. & Clark, J. W., The US Code. *Proc. Symp. Aluminium in Structural Engineering*. Institution of Structural Engineers, London, 1963.

Hill, H. N., Hartmann, E. C. & Clark, J. W., Design of aluminium alloy beam columns. *Trans. ASCE*, **121** (1956) 1.

Hill, H. N., Clark, J. W. & Brungraber, R. J., Design of welded aluminium structures. *Proc. ASCE* (June 1960).

Hong, G. M., Aluminium column curves. *Proc. Int. Conf. on Steel and Aluminium Structures*, Cardiff, July 1987. Elsevier Applied Science, London.

Institution of Structural Engineers, Report on the structural use of aluminium, January 1962, London.
Institution of Structural Engineers. Symposium on the structural use of aluminium, October 1985, London.
Kloppel, K. & Barsch, W., Versucke zum Kapitel Stabilitatsfalle der Neufassung von DIN 4113. *Aluminium,* Heft 10 (1973) 690.
Kosteas, D., Welded aluminium components—fatigue behaviour, design considerations and interrelation to quality recommendations. *Proc. Int. Conf. on Quality and Reliability in Welding,* Vol. 3, paper C16, Hangzhou, China, September 1984.
Kosteas, D., Aluminium beam test program—assessing the results. Int. Conf. TWI Fatigue of Welded Constructions, Brighton, 1987.
Kosteas, D. & Sanders, W. W., Evaluation of fatigue strength for common welded joints in aluminium alloys. *Proc. 3rd Int. Conf. on Aluminium Weldments,* Munich 1985. Aluminium Verlag, Dusseldorf, 1985.
Kosteas, D. & Poalas, K., Considering reliability in fatigue design values estimation. *Proc. Int. Conf. on Steel and Aluminium Structures,* Cardiff, July 1987. Elsevier Applied Science, London, p. 145.
Kosteas, D. & Sanders, W. W., Fatigue design format for welded aluminium structures. *Proc. Int. Conf. on Steel and Aluminium Structures,* Cardiff, July 1987. Elsevier Applied Science, London, p. 156.
Lincoln, J. W., Damage tolerance—USAF experience. *Proc. 13th Symposium of the Int. Committee on Aero Fatigue,* Pisa, (1985) 265–95.
Little, G. H., Collapse behaviour of aluminium plates. *International Journal of Mechanical Science, London,* **24** (1) (1982) 37.
Lyst, J. O., Effects of stretching on the fatigue strength of 2024-T4 aluminium alloy. *Materials, Research and Standards,* September 1962.
Maddox, S. J., British fatigue design rules for welded aluminium. *Proc. Second Int. Conf. on Aluminium Weldments,* Munich, 1982. Aluminium Verlag, Dusseldorf.
Maddox, S. J., The effect of tensile residual stresses on the fatigue strength of transverse fillet welded AlZnMg alloy. Contract report for RARDE (Christchurch), 1986.
Marshall, W. T., Nelson, H. M. & Smith, I. A., Experiments on single-angle aluminium alloy struts. *Proc. Symp. Aluminium in Structural Engineering.* Institution of Structural Engineers, London, 1963.
Massonnet, C. & Hagon, R., The draft Belgian standard for the use of aluminium in structures. *Proc. Symp. Aluminium in Structural Engineering.* Institution of Structural Engineers, London, 1963.
Mazzolani, F. M., *Aluminium Alloy Structures.* Pitman Books International, London, 1985.
Mazzolani, F. M. & Faella, C., European buckling curves for aluminium alloy welded members. *Alluminio,* No. 10 (1980).
Mazzolani, F. M., Cappelli, M. & Spasiano, G., Plastic analysis of aluminium alloy members in bending. *Aluminium,* **61** (1985).
McLester, R., Fatigue strength of welded and riveted joints in aluminium. *Proc. Symp. Aluminium in Structural Engineering.* Institution of Structural Engineers, London, 1963.

Mofflin, D. S., A finite strip method for the collapse analysis of compressed plates and plate assemblages. Cambridge University Report ED/D—Strut/TR100, 1983.

Mofflin, D. S. & Dwight, J. B., Buckling of aluminium plates in compression. In *Behaviour of Thin-walled Structures—1*, ed. J. Rhodes, Elsevier Applied Science, 1984.

Moore, R. L., Observations on the behaviour of aluminium alloy test girders. *Proc. ASCE*, **72** (1946) 729.

Muckle, W., *The Design of Aluminium Alloy Ships' Structures*. Hutchinson, London, 1963.

Nelson, H. M., Report on tests of aluminium alloy angles in tension. Aluminium Development Association Memorandum No SR/35, 1954.

Nethercot, D. A., Aspects of column design in the new UK structural aluminium code. *Proc. Int. Conf. on Steel and Aluminium Structures*, Cardiff, July 1987. Elsevier Applied Science, London.

Peel, C. J. *et al.*, The development and application of improved aluminium–lithium alloys. 2nd Int. Conf. on Al-lith Alloys, Monterey, California, April 1983.

Pohler, C. H. *et al.*, A technology base for aluminium ship structures. *Naval Engineers' Journal*, **91** (5) (1979).

Pook, L. P., Fracture mechanics; how it can help engineers. Trans N-E Coast Inst. Engineers and Shipbuilders, **90** (3) (1974) 77–92.

Ramirez, J. L., Aluminium structural connections: conventional slip factors in friction grip joints. *Proc. Int. Conf. on Steel and Aluminium Structures*, Cardiff, July 1987. Elsevier Applied Science, London, p.115.

Ramirez, J. L., Mendia, J. & Zalbidea, J. A., Verification of the conventional friction coefficient on high strength preloaded bolted connections in aluminium structures. Escuela Superior de Ingenieros Industriales de Bilbao, ECCS T2-55-81, May 1981.

Ramirez, J. L., Zalbidea, J. A. & Feijoo, J. M., Additional results about conventional friction coefficient in preloaded bolted connections in aluminium structures. Escuela Superior de Ingenieros Industrailes de Bilbao, February 1984.

Redshaw, S. C., Aluminium angles in tension. *Proc. Symp. Aluminium in Structural Engineering*. Institution of Structural Engineers, London, 1963.

Rockey, K. C., Aluminium plate girders. *Proc. Symp. Aluminium in Structural Engineering*. Institution of Structural Engineers, London, 1963.

Rockey, K. C., Evans, H. R. & Porter, D. M., A design method for predicting the collapse behaviour of plate girders., *Proc. ICE*, **65** (March 1978).

Rühl, K., Development of Standards for the use of aluminium in building construction. *Zeitschrift des Veriens Deutscher Ingenieure*, **102** (1960) 1712.

Rühl, K., Review of German Standards for aluminium in structural engineering. *Proc. Symp. Aluminium in Structural Engineering*. Institution of Structural Engineers, London, 1963.

Schijve, J., Fundamental and practical aspects of crack growth under corrosion and fatigue conditions. *Proc. Inst. Mech. Engrs*, **191** (1977) 107–14.

Sedlacek, H. & Sedlacek, G., Die S-Brücke-Ein leistungsfähiges Pionierbückengerät aus Aluminium. *Aluminium*, Heft 4 (1977) 260.
Sedlacek, G. & Ungermann, D., Dimple plates for light weight structures in steel and aluminium. *Proc. Int. Conf. on Steel and Aluminium Structures*, Cardiff, July 1987, Elsevier Applied Science, London, p. 93.
Smith, C. S., Compressive strength of welded steel ship grillages. NCRE report R 611, May 1975.
Smith, R. E., Column tests on some proposed aluminium standard structural sections. Aluminium Development Association Research Report No. 28, 1955.
Soetens, F., Welded connections in aluminium alloy structure. Second Int. Conf. on Aluminium Weldments, Munich. Aluminium-Verlag, Dusseldorf, 1982.
Soetens, F., Connections in aluminium alloy structures. *Proc. Int. Conf. on Steel and Aluminium Structures*, Cardiff, July 1987. Elsevier Applied Science, London, p. 105.
Steinhardt, O., Connections for aluminium constructions in structural engineering. *Zeitschrift des Veriens Deutscher Ingenieure*, **102** (1960) 1729.
Stussi, F., On the fatigue of metals with special reference to aluminium. *Proc. Symp. Aluminium in Structural Engineering*. Institution of Structural Engineers, London, 1963.
Subcritical crack growth. Working Group of the Fracture Sub-Committee of the Royal Society—British National Committee on Theoretical and Applied Mechanics, London, Nov. 1984.
Sumpter, J. D. G. *et al.*, Fracture toughness of ship steels. *Roy. Inst. Naval Arch., Spring Meeting*, paper 2, (1988) 1–9.
Tomlinson, J. E. & Wood, J. L., Factors influencing the fatigue behaviour of welded aluminium. *British Welding Journal* (April 1960).
Valtinat, G., Untersuchung zur Festlegung zulässiger spannungen im Kräfte Bei Niet-Bolzen-und HV-Verbindungen aus Aluminium legierungen. *Aluminium* **47** (1971).
Valtinat, G. & Dangelmaier, P., Näherungsweise Berechnung der Traglasten von nichtgeschweissten und geschweissten Druckstäben aus Aluminium. *Der Bauingenieur*, **61** (1986) 507.
Vilnay, O., Behaviour of aluminium alloy struts and beams reinforced by adhesive bonding of carbon fibre/epoxy composites. *Proc. Int. Conf. on Steel and Aluminium Structures*, Cardiff, July 1987. Elsevier Applied Science, London, p. 135.
Webber, D., The effect of transverse fillet welds on the fatigue life of DGFVE 232A (7019) AlZnMg alloy. International Institute of Welding, Boston Conference, 1984.
Webber, D., Strength of welded 232 AlZnMg alloy. RARDE (Christchurch), Branch Note EE/BR101, 1983.
Webber, D., Strength of DGFVE 232A AlZnMg alloy, artificially aged after welding. RARDE (Christchurch), Branch Note EE/BR 106, 1983.
Webber, D., Validation of design properties of welded DGFVE 232A AlZnMg alloy, naturally aged after welding, RARDE (Christchurch), Branch Note EE/Br 68, 1983.

Webber, D., Statistical analysis of DGFVE 232 AlZnMg alloy properties. RARDE (Christchruch), Branch Note EE/BR 102, 1983.

Webber, D. & Maddox, S. J., Fatigue crack propagation in AlZnMg alloy fillet welded joints. American Society for Testing and Materials, STP 648, 1978.

Webber, D. & McDiarmid, D. L., A comparison of various welded aluminium alloy joint specimens in tenstile fatigue at short lives. International Institute of Welding, Boston Conference, 1984.

Weinhold, J., Load stresses of compression bending members made of aluminium. *Aluminium,* **34** (3) (1958).

Wong, M. P., Weld shrinkage in non-linear materials. PhD thesis, Cambridge University, September 1982.

2

Bars, Beams and Beam Columns

F. M. Mazzolani
Università Federico II,
Naples, Italy

&

G. Valtinat
Technische Universität,
Hamburg-Harburg, Germany

assisted by

A. Dé Luca
A. De Martino
C. Faella

&

P. Dangelmaier

ABSTRACT

The structural aluminium version of the 'industrial bar', first introduced to help deal with the problem of geometrical and mechanical imperfections in steel structures, is discussed. This is followed by a detailed study of the strength of aluminium elements in tension, bending and compression. The ultimate behaviour of beams in bending, the effects of welding, and the problem of buckling in thin sections are all reviewed.

The design of columns and beam columns, including a thorough analysis of buckling behaviour, is included, with particular emphasis on computer simulation. The work described here was linked mainly to the production of an aluminium design document for the European Convention for Constructional Steelwork.

1 DEFINITION OF AN 'INDUSTRIAL BAR'

1.1 General

Of the several difficulties that are encountered in the theoretical analysis of static and stability problems for aluminium alloy structures, the first is related to the idealization of the tensile material properties.

The alloys commonly used in structures possess mechanical properties considerably different from each other. Even properties of the same alloy differ when it undergoes different fabrication processes and heat treatments.

Furthermore, the continuous behaviour of the σ–ε curve, obtainable in a tension test, cannot be simplified to an elastic/perfectly plastic behaviour as is done for mild steel. More complex models have to be used, by considering a material with a generalized inelastic behaviour (see Section 1.2).

The effects of unavoidable imperfections due to the fabrication process were introduced in the design of steel structures during the 1960s. Starting from 1970, research activity widely developed by the ECCS Committee on Aluminium Alloy Structures led for the first time to a definition of what 'industrial bar' means for aluminium alloys, both extruded and welded, and the main differences with steel were emphasized.

As for steel, the aluminium alloy 'industrial bar' possesses geometrical and mechanical imperfections which actually influence its load bearing capacity. These are as follows:

Geometrical imperfections which are usually understood to be the differences between the nominal and the actual geometry of the structural element, for the longitudinal dimensions as well as for the transverse dimensions of the cross-section (see Section 1.3).

Mechanical imperfections which can be grouped into two different types:

the residual stress distribution (see Section 1.4.1);

the inhomogeneous distribution of mechanical properties (see Section 1.4.2).

These subjects are widely developed in Ref. 1.

1.2 The stress–strain relationship

1.2.1 Piecewise idealization

The simplest way to model the stress–strain relationship of aluminium alloys is with two straight lines representing an elastic-hardening diagram (see Fig. 1(a)). The first part of the diagram, which represents the elastic portion, starts from the origin with a slope equal to the Young's modulus E_o. The second part of the diagram, which represents the strain-hardening portion, has a slope equal to the tangent modulus E_1. The intersection of the two lines defines the conventional value f_p of the elastic limit of proportionality.

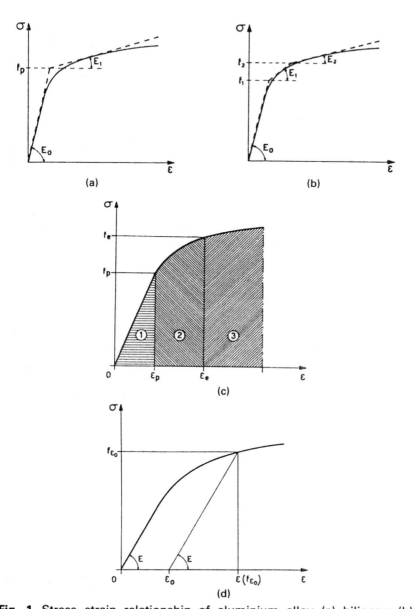

Fig. 1. Stress–strain relationship of aluminium alloy (a) bilinear; (b) trilinear; (c) three separate portions of the material behaviour; (d) definition of ε_0 and $\varepsilon(f_{\varepsilon_0})$.

This type of approach can be improved if an intermediate line which is a tangent to the actual curve at the 'knee' is introduced (see Fig. 1(b)). The polygon which derives from this idealization is therefore characterized by three moduli E_0, E_1 and E_2, and by two reference stresses f_1 and f_2.

1.2.2 Continuous models of form $\sigma = \sigma(\varepsilon)$

It is very difficult for a law of the form $\sigma = \sigma(\varepsilon)$ to be both very general and very close to the actual inelastic behaviour of the material. In any case, to be utilized in design, whatever expression is used for the σ–ε law has to be related to the conventional experimental values such as $f_{0.1}$, $f_{0.2}$ and E.

It is usually convenient to identify three separate portions of the function $\sigma = \sigma(\varepsilon)$, in each of which the behaviour of the material differs. They can be defined in the following way (Fig. 1(c)):

Region 1   elastic behaviour
Region 2   inelastic behaviour
Region 3   strain-hardening behaviour

In each region, the σ–ε relationship which represents the behaviour has to be found. To ensure continuity the three different laws have to produce coincident points and slopes at their intersection.

This methodology has been followed by Baehre,[2] whose studies have been applied in the Swedish specifications, and by Mazzolani[3] during the research carried out by the Committee on Light Alloy Structures of the ECCS.

1.2.3 Continuous models of the form $\varepsilon = \varepsilon(\sigma)$: Ramberg–Osgood law

A generalized law $\varepsilon = \varepsilon(\sigma)$ has been proposed by Ramberg & Osgood[4] for aluminium alloys:

$$\varepsilon = \frac{\sigma}{E} + \left(\frac{\sigma}{B}\right)^n \qquad (1)$$

where E is the Young's modulus at the origin, and B and n have to be determined by experiment.

The physical meaning of B and n can be expressed as follows (Fig. 1(d)). If an elastic limit stress is defined such that a residual strain will be left when the specimen is unloaded, eqn (1) for $\sigma = f_{\varepsilon_0}$ will be:

$$\varepsilon = \frac{f_{\varepsilon_0}}{E} + \left(\frac{f_{\varepsilon_0}}{B}\right)^n \qquad (2)$$

We also have

$$\frac{f_{\varepsilon_0}}{E} = \varepsilon - \varepsilon_0 \tag{3}$$

By substituting eqn (2) into eqn (3) we obtain

$$\varepsilon_0 = \left(\frac{f_{\varepsilon_0}}{B}\right)^n \tag{4}$$

and therefore

$$f_{\varepsilon_0} = B\sqrt[n]{\varepsilon_0} \tag{5}$$

If we assume $\varepsilon_0 = 0.002$, then

$$f_{0.2} = B\sqrt[n]{0.002} \tag{6}$$

Alternatively, if we assume $\varepsilon_0 = 0.001$, then:

$$f_{0.1} = B\sqrt[n]{0.001} \tag{7}$$

From the ratio between eqn (6) and eqn (7) we obtain:

$$\frac{f_{0.2}}{f_{0.1}} = \sqrt[n]{2} \tag{8}$$

Equation (8) relates the exponent n to the strain-hardening parameter $f_{0.2}/f_{0.1}$ and hence, according to Sutter, to the heat treatment of the material. The exponent n of the Ramberg–Osgood law is therefore characteristic of the strain-hardening rate of the inelastic portion of the σ–ε diagram. It can be expressed as:

$$n = \frac{\ln 2}{\ln\left(\dfrac{f_{0.2}}{f_{0.1}}\right)} \tag{9}$$

When the ratio $f_{0.2}/f_{0.1}$ tends to 1, the value of n tends to infinity, and eqn (1) then represents the behaviour of mild steels. In this case we have:

$$\varepsilon = \frac{\sigma}{E} + \left(\frac{\sigma}{B}\right)^\infty \tag{10}$$

which gives:

$$\varepsilon = \frac{\sigma}{E} \quad \text{for} \quad \frac{\sigma}{B} < 1 \text{ (perfectly elastic portion)}$$

$$\varepsilon = \infty \quad \text{for} \quad \frac{\sigma}{B} > 1 \text{ (perfectly plastic portion)}$$

The two regions are separated by the value $\sigma/B = 1$, which corresponds to the knee of the elastic/perfectly plastic $\sigma-\varepsilon$ diagram typical of mild steels. The parameter B has the physical meaning of the limit stress of the elastic part of the curve when $n = \infty$. More generally it can be said that, for a finite value of n, the parameter B shows the extent of the curve for which the first term (σ/ε) of the Ramberg–Osgood law is more significant than the second $(\sigma/B)^n$. Furthermore, the ratio $f_{0\cdot2}/f_{0\cdot1}$ tends to 1 in eqn (8) when $n \to \infty$, whereas for $n = 1$ eqn (1) becomes linear and the ratio $f_{0\cdot2}/f_{0\cdot1}$ is equal to 2. It is therefore interesting to compare the values of the exponent n of the Ramberg–Osgood law and the values of the ratio $f_{0\cdot2}/f_{0\cdot1}$ which gives the thresholds for the Sutter classes (see Fig. 2(a) and Table 1).

In this classification,[5] the behaviour of the $\sigma-\varepsilon$ law is related to the heat treatment of the alloy on the basis of the values of the

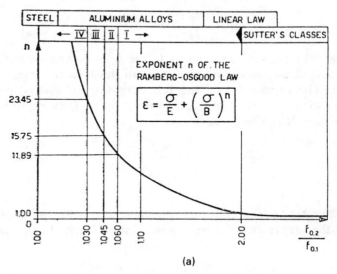

Fig. 2. Relation between exponent n of the Ramberg–Osgood law and the ratio $f_{0\cdot2}/f_{0\cdot1}$. (a) Sutter's classes.

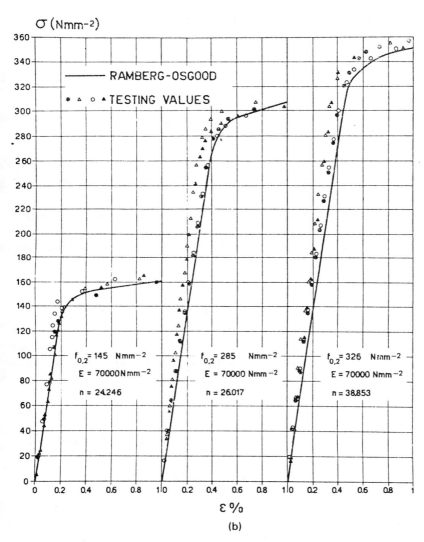

Fig. 2—*contd.* (b) Experimental data versus Ramberg–Osgood law.

Table 1. Relation Between Exponent n of the Ramberg–Osgood Law and the Ratio $f_{0.2}/f_{0.1}$ (see also Fig. 2(a))

		$f_{0.2}/f_{0.1}$	n
Linear law		2	1
Sutter	I class	1·060	11·89
	II class	1·045	15·75
	III class	1·030	23·45
IV class		1	∞
Mild steel			

strain-hardening parameter. Three principal classes are established:

Class I   annealed alloys (if $f_{0.2}/f_{0.1}$ is greater than 1·060),
Class II tempered alloys (if $f_{0.2}/f_{0.1}$ falls between 1·045 and 1·060),
Class III quenched and tempered alloys (if $f_{0.2}/f_{0.1}$ falls between 1·030 and 1·045).

This classification is based upon the assumption that there are no aluminum alloys with a value of $f_{0.2}/f_{0.1}$ less than 1·030. In reality, it has been shown during testing carried out by the ECCS Committee on Light Alloy Structures in 1972 that aluminium alloys with a value of $f_{0.2}/f_{0.1}$ less than 1·030 can be found. This can be explained by the fact that zinc and magnesium alloys have been developed since the Sutter classification was proposed.

It therefore seems logical to add a fourth class, which comprises those alloys with a value of $f_{0.2}/f_{0.1}$ less than 1·030. This class takes account of the most recent alloys.

It is possible to classify aluminium alloys according to the values of the exponent n of the Ramberg–Osgood law:

$$n < 10\text{--}20 \text{ (non-heat-treated alloys)}$$

$$n > 20\text{--}40 \text{ (heat-treated alloys)}$$

This has also been confirmed by experimental results.[6]

Equation (1) can be simplified by substituting B from eqn (6):

$$B = \frac{f_{0.2}}{\sqrt[n]{0.002}} \tag{11}$$

We obtain:

$$\varepsilon = \frac{\sigma}{E} + 0.002\left(\frac{\sigma}{f_{0.2}}\right)^n \tag{12}$$

The exponent n is a function of $f_{0.2}$ and $f_{0.1}$, and hence this form of the law is in terms of parameters that can all be determined experimentally from a tension test.

The Ramberg–Osgood law is now widely used because its predicted behaviour is very close to the actual behaviour of aluminium alloys.

Figure 2(b) shows some comparisons for different non-heat-treated and heat-treated alloys. Ramberg–Osgood curves give a lower bound to the experimental curves owing to the use of the minimum value of Young's modulus ($E = 70\,000\,\text{N mm}^{-2}$).

For these reasons the Ramberg–Osgood law has been used in the computations carried out by the ECCS Committee on Light Alloy Structures when developing the European recommendations for aluminium alloy structures (1978).

A very simple proposal[7] is to assume $10n = f_{0.2}$ (N mm^{-2}).

1.3 Geometrical imperfections

1.3.1 Out-of-straightness

Industrial bars are never perfectly straight; they possess an initial out-of-straightness. This deformation can be idealized by a sinusoidal or parabolic expression which is characterized by the parameter v_o (displacement at midspan). This displacement is commonly expressed as a percentage of the total length of the structural component (Fig. 3(a)). A systematic analysis carried out on extruded profiles from several European countries showed that the difference between an industrial bar and an ideal straight bar is equal to about $L/2000$.[8]

This imperfection is usually higher in steel profiles which have been straightened by rolling. In these profiles a value of $v_o = L/1000$ is universally assumed in order to compute the load-bearing capacity of the bar under compression.

Extruded aluminium profiles are straighter because of the more severe tractioning process. However, national specifications are usually less restrictive since they accept v_o/L ratios equal to $1/500$–$1/1000$.

Further measurements of dimensions carried out on welded double T and box sections always gave v_o/L less than $1/1300$. Some results obtained on box sections, which gave the highest deviations from linearity, are given in Fig. 3(c).

In the light of the results obtained directly and of the specifications which control the fabrication processes in different European coun-

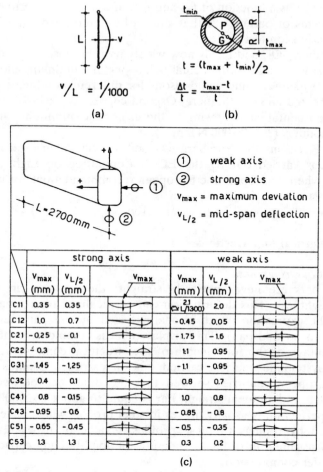

Fig. 3. Out-of-straightness. (a) Curvature. (b) Tubes. (c) Hollow sections.

tries, the ECCS committee decided to compute the instability curves given in its recommendations on the basis of an initial sinusoidal curvature. This takes $v_o/L = 1/1000$, thus being on the conservative side and in accordance with steel structures.

1.3.2 Variation of the dimensions of the cross-section

Measurements taken on some extruded profiles produced in several European countries showed that the dimensions of the cross-section

P				T				C		
	mean (mm)	+%	-%		mean (mm)	+%	-%		mean (mm)	+%	-%
h	195.09	0.88	-0.66	h	199.74	0.18	-0.22	h	196.61	0.35	-0.46
b	100.17	0.22	-0.27	h_f	198.13	0.59	-0.67	b	140.05	0.25	-0.25
t	11.95	0.38	-0.71	b	100.21	0.58	-0.61	b_m	139.52	0.55	-0.66
e	8.29	1.22	-0.71	t	12.00	2.45	-1.71	t	7.90	0.57	-0.94
				e	8.34	0.72	-1.67	e	5.13	1.24	-2.26

(d)

e_w = web eccentricity (mm)
e = load eccentricity (mm)

P	e_w	e	T	e_w	e
P11	-0.40	-0.10	T11	-0.13	-0.05
P12	-0.45	-0.11	T12	1.54	0.60 ≃ L/1800
P14	0.65	0.15	T14	-0.23	-0.09
P21	-0.33	-0.08	T21	0.04	0.02
P22	-0.75	-0.18 ≃ L/6000	T22	0.02	0.01
P24	0.43	0.10	T24	1.73	0.67 ≃ L/1600
P31	-0.03	-0.01	T31	0.	0.
P32	-0.28	-0.07	T32	-0.15	-0.06
P33	0.38	0.09	T33	0.26	0.10
P41	-0.65	-0.15	T41	0.34	0.13
P42	0.35	0.08	T42	-0.68	-0.26 ≃ L/4200

(e)

Fig. 3.—*contd.* (d) I-sections. (e) I-sections, web eccentricity.

(depth, width and thickness) are very close to the nominal dimensions. Different national specifications allow an average tolerance equal to 1% on general dimensions and 5% on the thickness of the different parts which form the open section. This value can reach 10% in the case of thin profiles less than 5 mm thick.

In the case of hollow sections the dimensions are less precise and national specifications allow higher tolerances. This has also been proved by dimensional measurements carried out on European tubes. This fact can be explained by the extrusion process of hollow sections which produces nonuniform thickness.

This imperfection, called 'eccentricity' by the ECCS committee, can be characterized by the ratio between the highest deviation of the thickness Δt and the average thickness t (see Fig. 3(b):

$$\frac{\Delta t}{t} = \frac{t_{max} - t}{t}$$

According to specifications, this ratio cannot exceed 10%. Measurements gave 'eccentricities' lying between 3 and 9%.

The most important consequence of this imperfection is to produce an initial eccentricity of the load applied at the ends of the bar. This is due to the fact that the load acts on the centre of the external perimeter, which is not coincident with the centre of gravity G (Fig. 3(b)). The negative effect of this type of eccentricity on the load-bearing capacity of columns has to be superimposed on that due to the initial out-of-straightness. This prediction has been confirmed by buckling tests on tubes which gave experimental values for the failure load of columns lower than those computed by the numerical analysis for the ideal tube.[8]

Measurements of the dimensions of welded profiles[9] showed that the deviations from the nominal values were always within the limits of tolerance allowed for extruded profiles (see Fig. 3(d)).

An important imperfection in welded double T profiles is the eccentricity of the web in the direction of the weak axis bending (Fig. 3(e)). This imperfection, even if within the tolerance limits, causes an eccentricity of the load which has to be added to the initial out-of-straightness.

As would be expected, this imperfection is usually larger in those profiles in which the web is directly welded to the flange by fillet welds (type T) than in those profiles whose flanges are extruded and butt welded to the web (type P). However, even the most severe condition, $e = L/1600$, when added to the initial out-of-straightness v_o, is such that it is less than the limit value $L/1000$:

$$(e + v_o) \leq L/1000$$

This shows that the $L/1000$ value can also be conveniently used to take account of the latter imperfection.

Numerical analysis carried out by the ECCS showed that variations in the dimensions of the cross-section from the nominal values produce a percentage reduction in the load-bearing capacity of columns. This can be assumed to be of the same order of magnitude as the dimensional variation itself.[10]

It is, however, very difficult to take account of this effect from a probabilistic point of view, particularly when considering the interacting effect of other mechanical and geometrical imperfections.

1.4 Mechanical imperfections

1.4.1 Residual stresses

1.4.1.1 Extruded profiles. There are several physical reasons for predicting that thermal residual stresses caused by cooling in extruded aluminium alloy profiles are smaller than those in similar hot-rolled steel profiles. In order to prove these ideas experimentally, testing has been carried out on extruded profiles of different alloys using the sectioning method.[1]

Double T $63 \times 63 \times 4$ mm extruded profiles were first examined, the material being the French alloys A-GSM, A-SGM, A-U4G and A-Z5G. It was observed from these test results (Fig. 4(a)) that the distribution of residual stresses is very irregular and does not follow any law like that for steel structures. The highest values of residual stresses (end of flanges, centre of web) are not very high in compression (lower than 50 N mm^{-2}) and are even lower in tension.

It should also be noted that these values have been measured on the surface of the profiles. At the centre of the material the values are probably lower, especially when, as is usual, residual stresses change sign from one side of the profile to the other.

The mechanical properties of the material did not affect the intensity and distribution of residual stresses.

Austrian profiles were then examined (Fig. 4). These consisted of an asymmetrical double T profile (b) and hollow sections of constant thickness with rectangular (c) and square (d) cross-section, made of 6060 and 7020 alloys.[11]

The aim of the testing was to identify the influence of the usual manufacturing phases on the formation of residual stresses.

Measurements were taken on profiles produced using different manufacturing processes:

(1) extruded and cooled by air,
(2) extruded and straightened (about 1%),
(3) extruded, straightened and artificially tempered (100°C for 4 h and 140°C for 24 h).

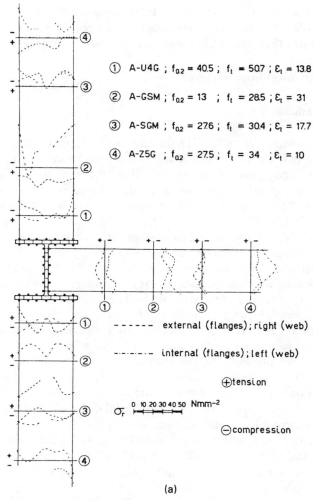

Fig. 4. Residual stresses, test results. (a) I-sections.

The results are not easily explained, owing to the low intensity of the initial straightening (1%) which did not lead to significant relief of residual stresses. However, final values confirmed that residual stresses produced by manufacturing are generally very low in extruded profiles (less than $20\,\mathrm{N\,mm^{-2}}$).

Very conservative values of residual stresses, obtained in symmetri-

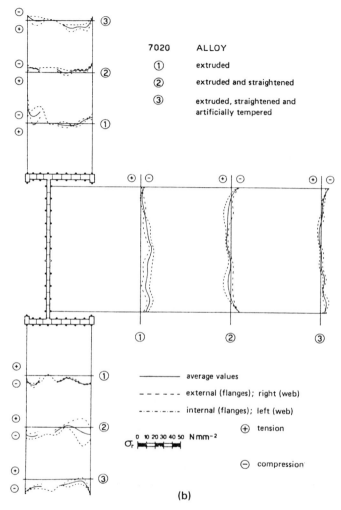

Fig. 4—*contd.* (b) Rectangular hollow section.

cal double T profiles, have been used to study their influence on instability. As might be expected, their effect was negligible. Thus in steel structures residual stresses due to cooling processes in hot-rolled profiles represent an important imperfection because they reach values approximately equal to 0·3–0·5 of the yield limit. By contrast, in

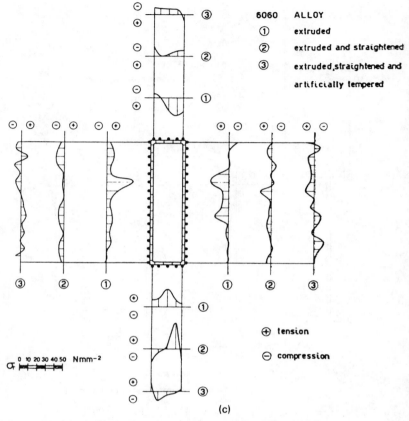

Fig. 4—*contd.* (c) Asymmetric I-section.

extruded aluminium alloy profiles, whatever the heat treatment, residual stresses have very small values; for practical purposes these have a negligible effect on load-bearing capacity.

1.4.1.2 Welded profiles. In contrast to what has been said for extruded profiles, residual stresses represent a mechanical imperfection which cannot be neglected in welded profiles. In fact these profiles are subjected to heat treatment which is very inhomogeneous. The welding process produces a concentrated heat input, the intensity of which is closely related to the type of procedure used—in particular to the weld sequence, the pass size and the depth of penetration of the weld.

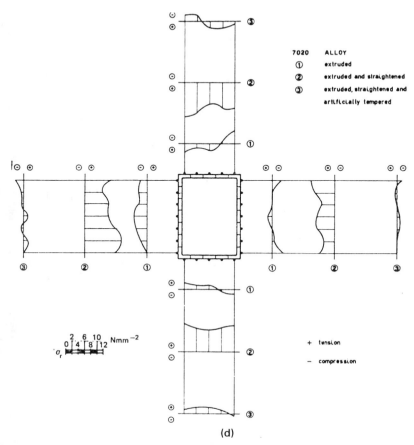

Fig. 4—*contd.* (d) Square hollow section.

The zones close to the welds are heated to very high temperatures and tend to expand, but this expansion is prevented by the regions further from the weld which are at lower temperatures. When the member has completely cooled, residual tension stresses close to the weld reach the yield limit, whereas equilibrating compression stresses arise further from the weld.

The intensity and distribution of residual stresses due to welding are always related to the thermal diffusion factor of the material, and are therefore lower than in steel structures.

Preliminary research has been carried out by Mazzolani[12] on simple groove butt-welded joints. Two French alloys were examined (Fig.

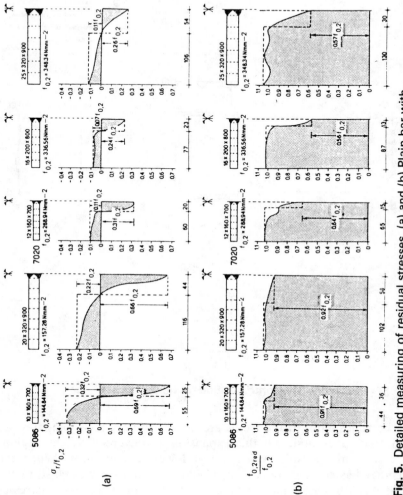

Fig. 5. Detailed measuring of residual stresses. (a) and (b) Plain bar with a weld.

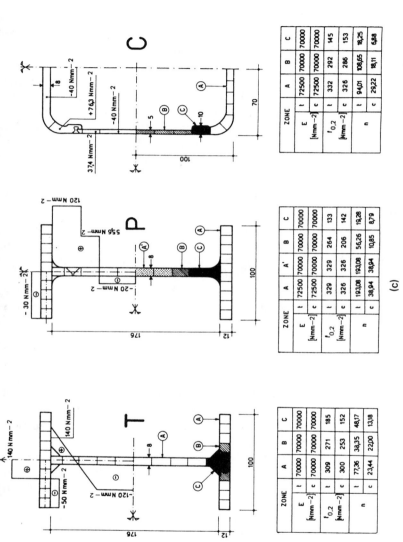

Fig. 5.—contd. (c) T-profile with welds, P-profile with welds and C-profile with welds.

5(a)): non-heat-treated alloy A-G4MCH (corresponding to AlMg4·5Mn), and heat-treated alloy A-Z5GT6 (corresponding to AlZnMg1). The welding material was A-65 (ALMg5).

Measurements taken using the sectioning method gave the results shown in Fig. 5(a), in which average residual stress distributions σ_r are drawn. The behaviour of the two materials is very similar. Residual stress values are very close to peak values, corresponding to the weld, of 100 N mm^{-2} and with equilibrating compression values of 30–50 N mm^{-2} at the end of the members. For a range of structural welded shapes, the results of the international research carried out by Gatto et al.[13] for the ECCS committee are available.

In this research three profiles in AlSiMg alloy (6082 type) have been examined. The results of the sectioning tests (Fig. 5(c)) showed that:

—The residual stress distribution is usually regular and is similar in profiles of the same shape.
—The highest values of residual stresses are tension stresses corresponding to the position of fillet welds, and they are equal to about 140 N mm^{-2}, which is the elastic limit of the weld metal.
—The highest compression values in the flanges are always smaller than 50 N mm^{-2}, whereas in the web they reach 120 N mm^{-2} in the double T shape with fillet welds and are smaller in the other two cases (20–40 N mm^{-2}).

These results confirm the theoretical estimates and the previous results on plate joints.

All these tests prove that residual stress distributions are characterized by tension regions close to the welds, where the highest values of tensile residual stresses are observed, and by equilibrating compression regions further from the welds. The highest values of compression residual stresses are always smaller than tension stresses.

The comparison with steel for a given structural profile (double T welded shape) in dimensionless form showed that in steel residual stresses in the weld can reach the yield limit of the parent metal. Even higher values occur because the weld metal usually has higher strength. In contrast to steel, the highest residual stresses in aluminium alloys, having the same strength as Fe360 (such as AlSiMg alloy), are less than 60% of the 0·2% elastic limit.

This favourable difference is also valid for compression stresses at the ends of the flanges, which in aluminium profiles reach about 20% of the reference stress, compared with 70% for steel.

Hence it can be said that the effect of residual stresses in lowering the resistance of welded compression bars in aluminium alloys is smaller than the corresponding effect in steel bars. However, this effect cannot be neglected as in extruded profiles, and must be taken account of in checking stability.

For this purpose it was necessary to convert experimental results into distribution models to be used as inputs to numerical simulations.[1]

1.4.2 Inhomogeneous distribution of mechanical properties

1.4.2.1 Extruded profiles. The distribution of the elastic limit along the cross-section of extruded profiles in aluminium alloys is quite uniform and is not closely related to the manufacturing process as for hot-rolled steel profiles. Some experimental results obtained at Liège showed that the greatest differences are no more than a few percent, and these are not significant with regard to load-bearing capacity of the members. This is in contrast to steel structures, in which greater differences occur.

This is the reason why the ECCS committee decided to ignore this imperfection in extruded profiles.

1.4.2.2 Welded profiles. Most aluminium alloys used for structural applications are heat treated or work hardened in order to improve their mechanical properties.

In welded profiles, the heat input removes some of the beneficial effects due to these treatments, and leads to a decrease in the elastic limit. The result is a distribution of strength varying along the cross-section of the profile, with a minimum at the welds. This imperfection is also related to a distribution of residual stress which, at the welds, is equal to the elastic limit of the annealed material.

The first experimental analyses carried out in the USA by Hill *et al.*[14] were on plate joints with longitudinal welds at the centre, using the 6000 series alloy. They led to the identification of a region close to the weld called the reduced-strength zone (Fig. 6(a)), the extent of which is equal to $2b_r$. This region corresponds to about one-half of the actual heat-affected zone, and a reduced, constant value, limit stress $f^*_{0.2}$ is attributed to this region.

As an upper bound to b_r, these researchers found a value of 0·74 in. On this basis, the French specifications (DTU régles Al, January 1971) give b_r values equal to or smaller than 25 mm.

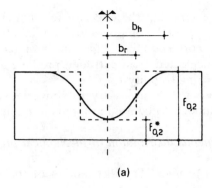

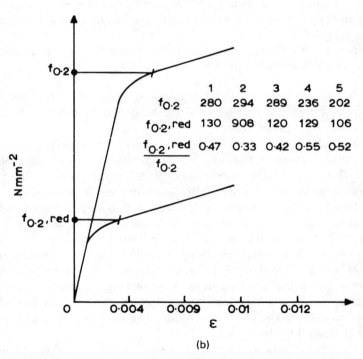

Fig. 6. Influence of welding. (a) Reduced strength zone. (b) Stress–strain law of base material and of material from the reduced strength zone.

The same researchers found the decreased values of the elastic limit $f^*_{0.2}$ given in Fig. 6(b). In comparison with the parent material, reductions of between 50 and 33% are observed.

This problem was later studied by Mazzolani from both theoretical[15] and experimental[12] points of view. Experiments carried out on the welded joints gave the average values given in Fig. 5(b) in which the variations of $f_{0.1}$ and $f_{0.2}$ are drawn along the different members.

As might be expected, the distributions depend greatly upon the heat treatment. Whereas for the non-heat-treated alloys (AlMg4·5Mn) the decrease of strength at the weld is about 10% for the heat-treated alloys (AlZnMg1) the decrease reaches 40–50%. The extent of the heat-affected zones confirmed the American results.

From tests carried out at the Experimental Institute for Light Metals (ISML) in Novara (1977)[13] on welded profiles of 6082 alloy it was observed that the distributions are usually homogeneous for each profile and have minimum values at the welds. These values correspond to the natural aging stage for the parent metal and to the annealed stage for the filler metal (AG5).

The extent of the decreased-strength region was always less than 20 mm on each side of the weld.

Three regions have been identified (Fig. 5(c)):

A   unaffected parent metal,
B   partially affected parent metal,
C   heat-affected zones around the weld metal.

The comparison between the values of $f_{0.2}$ in compression and in tension shows that their elastic limits do not substantially differ, except for case B. However, the behaviour of the curves in tension and in compression is completely different. In particular, after reaching the proportional limit the curves in compression join the knee more gradually than the curves in tension and they start at smaller values; in addition, the curves in tension are less gradual. Furthermore, it can be observed that in the range of small plastic deformations the compressive resistance is usually smaller than that in tension, whereas the contrary is true for larger deformations.

In 1985 Dangelmaier[16] published an extensive work in the field of detailed stress–strain laws of weld material and heat-affected material. The aim of this work was to get more information via special tension tests about the mechanical data of the material in relation to the distance to the centre of the weld, where more or less heat affect has

Table 2. Weld Preparation and Details on the Welding Process

Welding data	AlMgSil F 32 $t = 8$ mm	AlMgSil F 28 $t = 12$ mm
Welding process	MIG	
Size of butt weld	2 mm gap, 2 mm root face	
Number of runs	1	2
Filler material	S–AlMg5	
Preheating	No	
Intensity of current	200 A	220 A
Voltage	27 V	27 V
Welding wire	ϕ 1·6 mm	ϕ 1·6 mm
Feed of wire	9·6 cm/s	10·4 cm/s
Welding speed	0·43 m/min	0·43 m/min
Position of temperature measurements	0 1 2	0 1 2
Distance behind electrode (cm)	5·0 7·5 10·0	5·0 7·5 10
Temperature (°C)	500 180 55	450 230 55

taken place, and to correlate these results with other results coming from hardness tests and from the literature. The investigation was carried out with the material:

AlMgSil   F 32, plate thickness $t = 8$ mm,
AlMgSil   F 28, plate thickness $t = 12$ mm.

Two parts of a plate with edge preparation were butt welded together longitudinally. All welding data may be taken from Table 2. Additionally with a special gauge, the temperature behind the electrode has been measured in different places such as

—position 0:   in the centre of the weld,
—position 1:   at the free outer edge of the plate,
—position 2:   in the middle between the weld and the free edge of the plate.

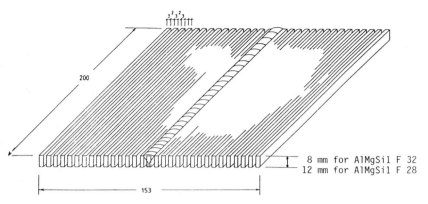

Fig. 7. Strip cutting of a plate parallel to the weld.

Although these measurements of the temperature could not be taken with full accuracy the results are presented in Fig. 7.

The 2·90 m long specimen (AlMgSi1 F 32, $t = 8$ mm) has been divided into 14 pieces, and the 2·50 m long specimen (AlMgSi1 F 28, $t = 12$ mm) into 12 pieces. These 200×200 mm test pieces have been cut into strips of 3 mm while the slit from the saw-blade was 2 mm (see Fig. 7). Thus it could be possible to have test pieces of n times 5 mm distance apart from the weld up to 50 mm and to achieve the corresponding $\sigma-\varepsilon$ laws. In total 546 deformation controlled tension tests have been performed, whose mean values of the $\sigma-\varepsilon$ curves, 95% confidence interval of the mean values and 5%- and 95%-fractile of all $\sigma-\varepsilon$ curves could be statistically evaluated. They are shown in the Fig. 8(a–f) for strips up to 30 mm apart from the weld. The figures contain also the Ramberg–Osgood laws for n-values which gave the best fit. Figures 9(a,b) summarize the results, they show the mechanical properties $f_{0·2,\mathrm{red}}$ of the weld and the heat affected zone material, and $f_{t,\mathrm{red}}$ of the welded and the heat affected zone material, as mean values.

With these data it is possible to allocate in later computer simulations to each finite element the corresponding stress–strain law depending on its distance to the centre of the weld. It was the optimum to distinguish between 8 distinct such curves M 1–M 8 where M 8 is valid for the weld and M 1 for the base material, while the others lie in between (see Table 3).

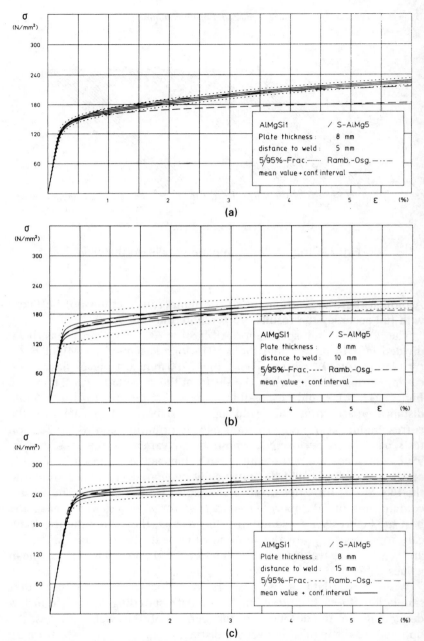

Fig. 8. Stress–strain laws of different strips in the distance, a, from the weld; plate thickness $t = 8$ mm. (a) $a = 5$ mm; (b) $a = 10$ mm; (c) $a = 15$ mm.

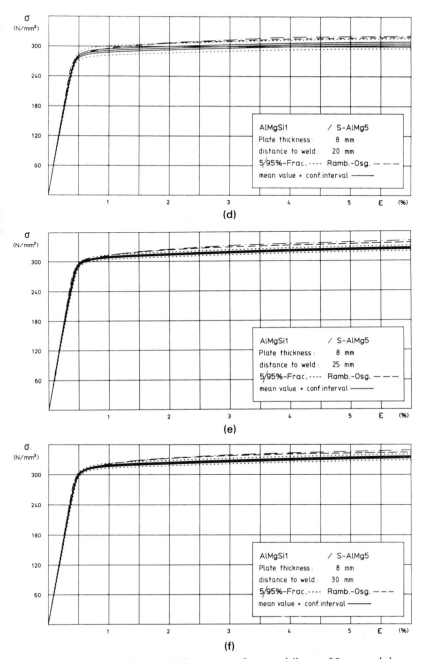

Fig. 8.—(Continued) Plate thickness t = 8 mm. (d) a = 20 mm; (e) a = 25 mm; (f) a = 30 mm.

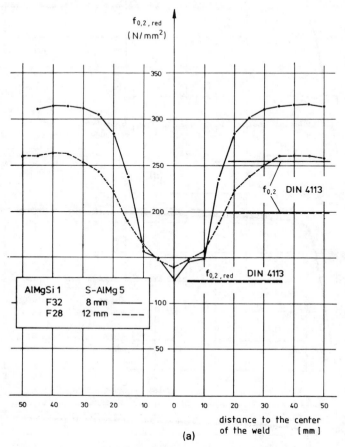

Fig. 9. Mechanical properties in the reduced strength zone (a) $f_{0.2, \text{red}}$.

2 MEMBERS IN TENSION

2.1 General

When a bar is subjected to simple tension, it reaches the ultimate limit state in the following cases:[1]

(a) When the ultimate strength f_t is reached in the gross cross-section of the bar.
(b) When rupture occurs in the connected part, i.e. in the net area of the bolted connection or in the reduced strength zones of the welded connection).

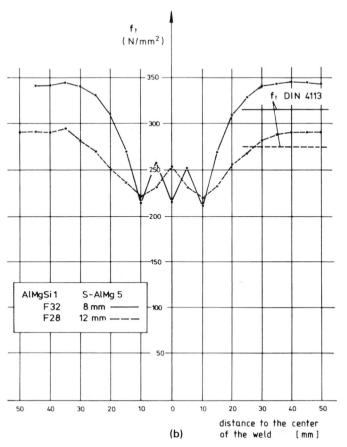

Fig. 9.—contd. (b) f_{ult}, red.

Table 3. Material Data of the Different Strips

Material No.	M8	M7	M6	M5	M4	M3	M2	M1
Distance to the centre of the weld (mm)	0	5	10	15	20	25	30	35–50
$f_{0.2}$ (N/mm^2)	125·1	147·4	150·8	238·8	284·8	303·3	311·9	314·1
E (N/mm^2)	67 057	69 307	71 448	71 110	70 982	70 616	69 618	71 145
F_t (N/mm^2)	186	205·0	180	268·4	307	326·5	336	337·9
ε_t (0/00) from tests	35	35	35	60	60	60	60	60
n	7·01	8·41	15·75	38·55	44·32	45·06	44·57	48·24

The first case corresponds to a ductile failure because the material can elongate in the inelastic hardening region before reaching the ultimate stress f_t (case 1, Fig. 10(a)).

The second case corresponds to failure of the connection, which can occur with different degrees of ductility depending upon the connection itself (cases 2, 3 and 4 of Fig. 10(a)). It is brittle in the case of bolted connections (case 3 of Fig. 10(a)) in which rupture occurs when the bar is still in the elastic range: these connections are therefore defined as 'partial restored strength'. Those connections in which the failure load of the connection is equal to that of the elastic limit of the connected bar are termed 'complete restored strength'.

Full capacity welded connections can be certainly considered as having complete restored strength in the case of non-heat-treated alloys, which are not affected by the heat input of the weld (no reduced-strength zones). Heat-treated alloy connections can have both complete or partial restored strength depending upon the extent of the reduced-strength zones.

Bolted connections can also be considered both complete or partial restored strength, depending on the net section, bearing and shear capacity of the belts.

Complete restoring of strength can be obtained, together with an elongation capacity, by connecting a sufficient number of bolts to all the parts of the cross-section (case 4 of Fig. 10(a)).

2.2 Strength of elements

The design force of a member in simple tension under axial load is given by

$$N_d = f_d A_{res}, \tag{13}$$

where f_d is the design strength and A_{res} is the resistant area of the cross-section. In the case of bolted connections

$$A_{res} = A_{net}$$

where A_{net} is the net cross-sectional area allowing for holes or cut-outs.

In special cases where the structure is designed according to the elastic theory and no plastic redistribution is needed, higher values of

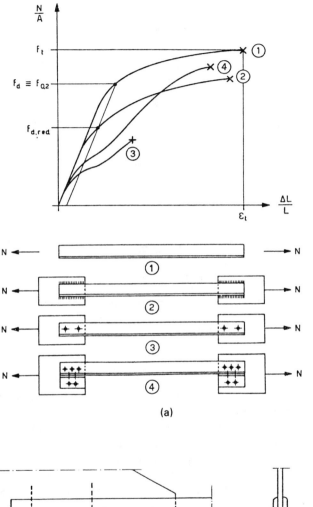

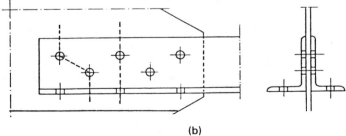

Fig. 10. Behaviour and ductility of bolted members. (a) Load deformation curve of different connections. (b) Staggered bolted connections.

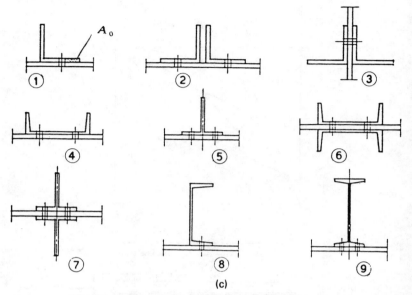

Fig. 10—*contd.* (c) Another type of connection.

N_d can be allowed, such as

$$N_d = \max \begin{cases} f_t A_{net}/1 \cdot 25 \\ f_{0 \cdot 2} A \end{cases} \quad (14)$$

In order to determine the minimum net area of a bar connected by means of bolts, various sections (straight, oblique and broken) are considered (Fig. 10(b)) and for each the length of surface traversed is summed up: the smallest of these multiplied by the thickness of material is adopted as the minimum net area.

In many cases the most unfavourable section is the one normal to the force and it is obtained by deducting the sum of empty spaces (holes) from the gross area. Various methods for evaluating net areas are proposed by codes (see Section 2.4).

In the case of welded connections, for members subject to HAZ softening

$$A_{res} = \left[A - \sum (1 - \beta) A_s \right] \quad (15)$$

where the summation is extended to all affected areas within the

Table 4. Reduction Factor β for the Relation Between the 0·2-Proof-Limits of HAZ Material and Base Material

Alloy	AlMg4·5Mn AlMg3 AlMg2Mn	AlZn4·5Mg1	ALMgSi1 ($f_t < 300$ N mm^{-2})	AlMgSi1 ($f_t \geqslant 300$ N mm^{-2})	AlMgSi0·5
β	1·00	0·65	0·60	0·50	0·50

cross-section; A_s is the section area of each softened zone; β is the reduction factor $= f_{0\cdot2\text{red}}/f_{0\cdot2}$. The reduced strength $f_{0\cdot2,\text{red}}$ is conventionally assumed equal to the elastic limit referred to the following stages (see Table 4):

- annealed (1*** and 5*** alloys);
- tempered in water and naturally aged (for 6*** alloys);
- tempered in the atmospheric and naturally aged (for 7*** series).

2.3 Ductility of connections

In cases where the structure is designed according to the plastic theory the following ductility requirements should be fulfilled:

The condition to be satisfied in order to have a complete restored bolted joint in a bar subjected to tension is

$$R_b \geqslant N_d \tag{16}$$

where R_b is the lesser of the values of the shear failure of the bolts and the bearing failure of the connected plate, and $N_d = f_{0\cdot2}A$ is the design strength of the bar (assuming $f_d = f_{0\cdot2}$ in eqn (13)).

The following equation has to be satisfied, together with eqn (16)

$$R_t \geqslant N_d \tag{17}$$

in order to guarantee failure in the section which is highly stressed by the bolted connection after yielding of the gross cross-section. R_t represents the collapse load of the net section and can be expressed as

$$R_t = \alpha f_t A_{\text{net}}, \tag{18}$$

where

f_t = ultimate stress of the material,
A_{net} = net area of the cross-section,

α = a coefficient less than 1 which takes into account the reduction in strength, with respect to the theoretical value, due to stress concentration along the holes and secondary stresses. This value has to be determined by tests.

We can assume $\alpha \simeq 1$ when the connection is designed, so that secondary stresses are eliminated (when the bar has double symmetry with respect to the gusset plate, thus eliminating eccentricities due to the shape of the cross-section between the axis of the bar and the centre of the reaction of the support). In the case of asymmetrical connections the value of α can be considered to lie between 0·8 and 0·9.

Conditions (16) and (17), which guarantee complete restoration of the resistant cross-section, can be written in the form:

$$\begin{cases} R_b \geq f_{0\cdot2} A \\ \alpha f_t A_{net} \geq f_{0\cdot2} A \end{cases}$$

The first equation implies that rupture occurs in the bar and not in the connection and the second provides plasticity in the bar before collapse of the net section. From the second equation the following is obtained:

$$\alpha \geq \frac{f_{0\cdot2}}{f_t} \frac{A}{A_{net}} \tag{19}$$

Whereas the A_{net}/A ratio in a bolted joint can be considered to lie between 0·8 and 0·9, the $f_t/f_{0\cdot2}$ ratio is dependent upon the type of alloy and varies between 1·1 and 2·5.

From eqn (19) the following classification is derived:

(a) For $f_t/f_{0\cdot2}$ ratios less than 1·25, bolted connections are unable to provide enough ductility typical of complete restored joints in plastic structures.
(b) For $f_t/f_{0\cdot2}$ ratios greater than 1·25 and less than 1·50, it is possible to obtain a complete restored joint provided that the secondary stresses are eliminated.
(c) For $f_t/f_{0\cdot2}$ ratios greater than 1·50, the connection is ductile enough to be considered as complete restored even though it is not symmetrical.

From the examination of commercial aluminum alloys it can be observed that: 5*** series alloys fall in categories (b) and (c), allowing complete restored connections: 2***, 6*** and 7*** series alloys usually fall in category (b), although in some cases they fall in category (a) and hence may not be suitable in ductile structures with bolted connections.

2.4 Codification

The new BS 8118 code (1985 draft) proposes to evaluate the net area A_{net} (14) as the smaller of the following cases:

—the full area A at a section perpendicular to the direction of loading less the areas of all holes or cut-outs at that section;
—the least area on any diagonal or zig-zag section through a chain of holes, given by

$$A_{net} = A - \sum A_h + \sum z^2 t / \Delta g \qquad (20)$$

where

z is the pitch of holes longitudinally,
g is the transverse spacing to hole centres,
t is the section thickness,
$\sum A_h$ is the sum of hole areas lying on the diagonal section considered.

The same code assumes the following value for f_d in (13)

$$f_d = 0 \cdot 5 (f_{0 \cdot 2} + f_u) \qquad (21)$$

but not greater than $1 \cdot 2 f_{0 \cdot 2}$, i.e. the average between the 0·2% proof stress and the tensile strength of the alloy.

Codes usually contain special provisions for angles in tension; the resistant area for evaluating the axial stress is defined with reference to the type of connection (Fig. 10(c)).

For an angle connected to both flanges, the resistant area coincides with the net area (14).

For an angle connected to only one flange (Fig. 10(C1)), the resistant area is expressed by:

—one bolt
$$A_{res} = 2 A_o \qquad (22)$$

—two or more bolts
$$A_{res} = A_1 + \xi A_2$$

where

A_o is the external area of the connected flange (see Fig. 10(C1)),
A_1 is net area of connected flange,
A_2 is area of non-connected flange,
$\xi = 3A_1/(3A_1 + A_2)$

Two identical angles connected to a joint plate in the flange plane (Fig. 10(C2)), have a resistant area expressed by:

$$A_{res} = 2A_1 + 2\xi A_2 \tag{23}$$

where A_1 and A_2 have the foregoing meaning and $\xi = 5A_1/(5A_1 + A_2)$.

For two angles connected along the web with at least two intermediate connections (Fig. 10(C3)), then:

$$A_{res} = 2A_{net} \tag{24}$$

where A_{net} is net area of each angle.

For channels connected through the web and perpendicular sections connected by their flanges (Figs 10(C4,5,6,7)) the same criteria can be adopted as in the case of angles.

It is inadvisable to adopt connections fastening only one flange of a channel or I-section (Fig. 10(C8,9)).

The above practical rules have been derived from experimental works on steel, but they are also usually adopted for aluminium alloys.

3 MEMBERS IN BENDING

3.1 General

The ultimate limit state of a beam can occur in different circumstances depending upon the geometry of the beam (the span L, the L/h ratio of the individual parts, etc.), the loading and support conditions and the type of connection.

In the case of beams with a compact cross-section, in which local buckling or flexural torsional buckling are not likely to occur, the beam experiences the inelastic range after reaching the limiting elastic moment $M_{0.2}$ until the ultimate moment M_u is reached (Fig. 11(a)). This moment cannot be defined (as it is for steel structures) as the full plastic moment. In fact, due to the hardening behaviour of the σ–ε law of aluminium alloys, a limiting curvature κ_{lim} has to be defined

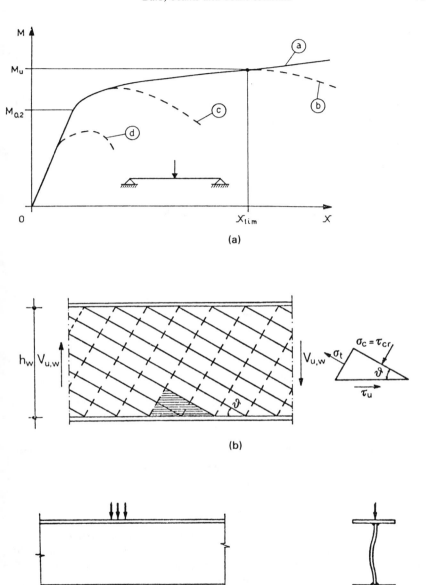

Fig. 11. Load carrying models. (a) Simple beam under concentrated load. (b) Diagonal stress field. (c) Local load with buckling effect.

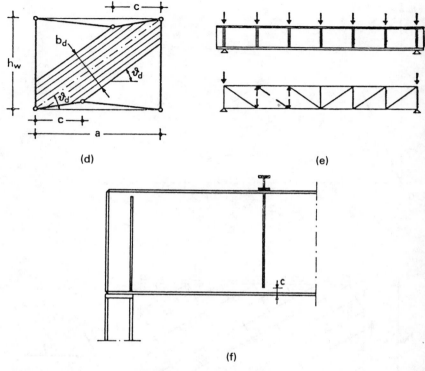

Fig. 11—*contd.* (d) Tension band action. (e) Beam with slender web—truss girder. (f) Stiffeners at concentrated load.

corresponding to the limit of large deformations in the inelastic range. The increase in strength from $M_{0.2}$ to M_u obtained in this phase can be quantified through a relation

$$M_u = \alpha M_{0.2}$$

which defines, in a generalized form, a new shape factor α which is not solely dependent upon the cross-sectional geometry, as is usual, but also depends upon the parameters of the σ–ε law ($f_{0.2}/E$ ratio, n exponent) and upon the definition of limiting curvature $\kappa_{\lim}$ (see Section 3.2.3).

In the case of open profiles, local buckling phenomena are most likely to occur in the compressed regions of the cross-section and cause a decrease in the M–κ curve of the beam. This unstable behaviour is dependent upon the b/t ratio (Fig. 11(a)). If the decreasing portion of the curve occurs after the ultimate moment M_u is reached (case b), the beam keeps the same maximum load carrying capacity. In this case the rotational capacity of the cross-section, which characterizes the flexural ductility of the beam, allows redistribution of the internal actions, and it is therefore possible to carry out a limit analysis of the whole structure (see Section 3.3).

If the decreasing portion of the curve occurs before the ultimate moment M_u is reached (case c), or even before the elastic moment $M_{0.2}$ (case d), the load carrying capacity of the beam is affected by local buckling phenomena, to a higher degree if the b/t ratio is large (e.g. thin profiles). Also ductility decreases to the extent that redistribution of internal actions cannot be considered.

In the case of slender beams (high L/h ratio) without transverse restraints, the ultimate limit state of the beam can be determined by flexural torsional buckling phenomena (see Section 3.4); these reduce the load carrying capacity of the beam by an amount dependent upon the values of slenderness.

Failure of a beam can also happen because of failure of an intermediate or edge connection. This failure will be brittle or ductile depending upon the connection, with complete or partial restored flexural resistance of the beam.

The required ductility of a connection depends upon its position along the beam. Since the maximum rotational capacity is required at the end of a beam, the edge connections (beam–beam connections or beam–column connections) must be sufficiently ductile to guarantee the redistribution of internal actions when the stresses are in the inelastic range $(M > M_{0.2})$, in order to reach the maximum load carrying capacity of the cross-section (M_u). Intermediate joints are used to connect the different sections of a beam constructed in more than one part. They usually correspond to the sections in which the bending moment is small (in close proximity to points of zero moment) and therefore do not need to exhibit complete restored flexural resistance.

Shear action usually occurs in conjunction with bending moment in beams and can cause collapse of the members, especially in the case of deep beams without stiffeners, in which the following unstable phenomena can occur:

(a) Local buckling of the web along the beam with consequent plane stress behaviour and formation of diagonal stress fields (Fig. 11(b)).

(b) Crippling of the web in positions corresponding to concentrated loads or support reactions (Fig. 11(c)).

When stiffeners are present, the structure can exhibit additional resistance in the post critical range.

From the constructional point of view, two systems can be identified:

(1) webs with transverse and longitudinal stiffeners,
(2) webs without intermediate stiffeners along the span.

In the first (traditional) system a critical state is characterized by local buckling of the web panel, with the flanges and the transverse stiffeners remaining rigid. This state can be identified provided that the appropriate values of extensional and flexural rigidities of the stiffeners are guaranteed. The critical state of the panel is idealized by diagonal bands of tensile stresses (Fig. 11(d)). Then, in the post critical range, the beam can be approximated by a triangulated structure in which the flanges represent the chords, the verticals are given by the transverse stiffeners and the bands of tension stresses represent the diagonals (Fig. 11(e)).

In these beams two limit states can be identified:

—A deformation limit state corresponding to local buckling of the panel.
—An ultimate limit state corresponding to collapse of the truss.

The first (deformation) limit state, which represents the maximum load before large transverse deformations of the panel, can be considered a serviceability limit state because local buckling of the panels can cause psychological and aesthetical inconvenience without compromising the load carrying capacity of the beam. The second (ultimate) limit state occurs because of the formation of a mechanism in the most stressed panel due to yielding in the tension band and plastic hinges in the flanges.

The second system, which seems to be more convenient on the basis of research carried out in the USA and Sweden, requires stiffeners only at supports and at positions of concentrated loads. The design of these beams, on the basis of steel specifications (AISC, Swedish code), guarantees that local buckling will not occur at positions of concentrated loads (Fig. 11(f)) even though it allows the web to buckle because of normal stresses due to shear and bending. The ultimate limit state is reached when the shear is given by

$$V_u = V_{u,w} + V_{u,f} \tag{25}$$

where

$V_{u,w}$ = ultimate value of shear due to plane stress conditions in the web,

$V_{u,f}$ = ultimate value of shear which the flanges resist because of their flexural rigidity.

The post critical behaviour which provides the $V_{u,w}$ value can be considered as corresponding to a system of orthogonal bars whose slope varies with the load. In fact, since there are no stiffeners, the truss model is no longer applicable.

3.2 Ultimate behaviour of cross-sections

3.2.1 Conventional elastic moment

(a) Extruded members. As a consequence of the definition of the elastic limit of the material, conventionally identified as the value of the stress which corresponds to a residual strain of 0·2% after unloading, a linear σ–ε law is assumed up to the $f_{0.2}$ value (Fig. 12(a)). If this limit is reached in the highly stressed fibres of a section subjected to bending, a limit state (referred to as 'conventional elastic') is identified, which corresponds to the elastic moment

$$M_{0.2} = f_{0.2} W \tag{26}$$

where W is the section modulus of the cross-section.

The moment $M_{0.2}$ conventionally provides the limit of the elastic behaviour. In fact, if we consider the real σ–ε law of the material, we observe that the limit of proportionality is smaller and that in the interval

$$f_p < \sigma < f_{0.2} \tag{27}$$

the stresses are in the inelastic range and cause residual deformations when the structure is unloaded (Fig. 12(a)).

Since the stresses are in the inelastic range, a residual midspan deflection v_r can be observed whose magnitude depends upon numerous factors such as the type of structure, the loading and support

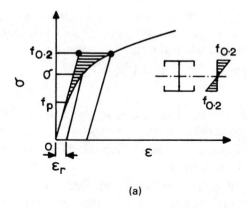

(a)

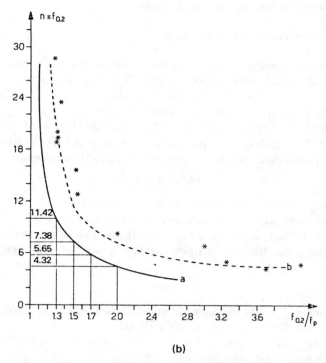

(b)

Fig. 12. Stress–strain–behaviour and moment–curvature–behaviours.

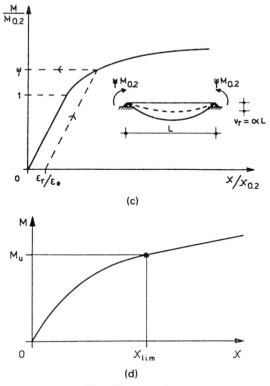

Fig. 12—contd.

conditions, the span length and cross-section of the beam and the type of alloy. Clearly the magnitude of this inelastic deformation depends upon the behaviour of the structural element and upon the extent of the $f_{0.2}$ to f_p range. If by convention it is assumed that

$$\varepsilon_p = 0.01\% \tag{28}$$

$$f_p = f_{0.01} \tag{29}$$

then from the Ramberg–Osgood law ((1)–(8))

$$f_p = f_{0.2}\sqrt[n]{0.05} \tag{30}$$

This relationship allows the calculation of $f_{0.2}/f_p$ ratios as functions of n. By assuming the coincidence $n = f_{0.2}$ (in $10\,\text{N/mm}^2$), curve (a) of Fig. 12(b) is obtained, showing a larger range for $f_{0.2}$ to f_p in non-heat-treated alloys (small values of n). This behaviour is also confirmed by some experimental results on Swiss alloys, interpreted by curve (b).

(b) Welded members. The conventional elastic moment of welded members can be evaluated by means of eqn (26) provided that the effects of the heat-affected zones are accounted for. In particular if the cross-section of the beam is built up by longitudinal welds, the geometrical properties of the reduced effective cross-section have to be used:

The reduced area:
$$A_{\text{red}} = A - \sum_{i=1}^{n} b_{\text{r},i} t_i [1 - f_{0\cdot 2,\text{red}}/f_d] \tag{31}$$

The reduced moment of inertia:
$$I_{\text{red}} = I - \sum_{i=1}^{n} b_{\text{r},i} t_i y_i^2 [1 - f_{0\cdot 2,\text{red}}/f_d] \tag{32}$$

The reduced section modulus:
$$W_{\text{red}} = I_{\text{red}}/y_{\text{max}} \tag{33}$$

where

A = nominal area of the cross-section normal to the axis of the weld (base metal and weld metal),
I = moment of inertia of the nominal cross-section of area A,
$b_{\text{r},i}$ = semiwidth of the single reduced strength zone,
t_i = average thickness of the base metal in that zone,
$f_{0\cdot 2,\text{red}}$ = reduced design strength in the reduced strength zone (see Section 2.2),
$f_{0\cdot 2}$ = design strength of the base metal,
y_i = distance between the centre of gravity in the ith reduced strength zone and the centre of gravity of the entire cross-section.

If all the cross-sections consist of transverse welds, it is sufficient to use the reduced design strength $f_{0\cdot 2,\text{red}}$ instead of the design strength $f_{0\cdot 2}$ of the base metal (see Section 2.2).

When transverse welds in a simply supported beam are at a distance from the ends less than $0\cdot 1L$, no reduction of strength is necessary.

3.2.2 *Plastic adaptation moment*

The classic plastic adaptation method introduced in the ECCS Steel Recommendations leads to results that are oversimplified and often unconservative and illogical for aluminium alloys. Even though this method is correctly used, the criterion of limiting the residual deformation after unloading to an assumed value (7·5% of the elastic deformation) does not take account of the geometry of the structure, which can be characterized in beams by the slenderness ratio L/h. On the basis of these considerations, and in order to eliminate the inconsistencies of the classical method, de Martino et al.[17] proposed a new plastic adaptation method based on the following criterion:

The plastic adaptation coefficient ψ can be calculated using the condition that the residual midspan deflection v_r due to bending is less than an assumed value αL, a function of the length L of the beam.

Consequently the residual deformation becomes independent of the material and is only related to the geometry of the beam through the slenderness parameter L/h. In fact, the position is

$$v_r = \alpha L \tag{34}$$

where, in the case of uniform bending,

$$\frac{v_r}{L} = \frac{\varepsilon_r}{4}\frac{L}{h} \tag{35}$$

and hence

$$\varepsilon_r = 4\alpha \frac{h}{L} \tag{36}$$

If α is taken as 1/1000, conventionally used to indicate initial values of the midspan deflection within tolerance limits, the following values of ε_r are obtained:

$\varepsilon_r = 0{\cdot}0004$ for $L/h = 10$
$\varepsilon_r = 0{\cdot}0002$ for $L/h = 20$
$\varepsilon_r = 0{\cdot}00013$ for $L/h = 30$

This corresponds to an approach which is more consistent than the previous definition based upon a constant percentage strain. The values of the computed deformation have the same meaning as the $\varepsilon_r = 0{\cdot}002$ strain conventionally used in defining the elastic limit of the material and do not require a plastic deformation proportional to the strength. This would be illogical because real materials in general become less ductile as strength increases.

The percentage value of residual deformation, which must be used instead of the classical value of 7·5% is given by

$$\frac{\varepsilon_r}{\varepsilon_{0\cdot 2}} = \frac{4E}{1000}\frac{h}{L}\frac{1}{f_{0\cdot 2}} \qquad (37)$$

and varies with the material $(E, f_{0\cdot 2})$ and the slenderness of the beam (L/h).

This method seems to be particularly applicable to families of materials with a large range of mechanical properties, such as aluminium alloys. For aluminium alloys characterized by $f_{0\cdot 2}$ in $10\,\text{N/mm}^2 \equiv n = 10$, 15, 20 and 30, and with slenderness ratios $L/h = 10$, 15, 20 and 30, and $\varepsilon_r/\varepsilon_{0\cdot 2}$ values are given as in Table 5. As can be seen, the deformation varies between 3·1 and 28% of the elastic deformation, depending upon the type of alloy. This variation is inversely proportional to the slenderness (L/h) and to the strength $(f_{0\cdot 2})$.

In order to use this method and to calculate the plastic adaptation coefficient ψ, the moment–curvature relationships are needed for the different cross-sections and for the alloys commonly used. These relationships can be obtained through a numerical procedure which discretizes the cross-section and uses small steps of the curvature. This procedure allows in this specific case a study of any type of cross-section, whether symmetrical, asymmetrical or more generally complicated shapes.

When the moment–curvature relationships are known, the values of the plastic adaptation coefficient ψ can be obtained by the intersection point between the moment–curvature diagrams with the unloading lines given by the following equation (Fig. 12(c)):

$$\frac{M}{M_{0\cdot 2}} = \frac{\kappa}{\kappa_{0\cdot 2}} - \frac{\kappa_r}{\kappa_{0\cdot 2}} = \frac{\kappa}{\kappa_{0\cdot 2}} - \frac{\varepsilon_r}{\varepsilon_{0\cdot 2}} \qquad (38)$$

Table 5. Limiting $\varepsilon_r/\varepsilon_{0\cdot 2}$-Values for the Plastic Adaptation Moment for Different Alloys and Different Slendernesses

$f_{0\cdot 2} \equiv n$ \ L/h	10	15	20	30
10	0·280	0·187	0·140	0·093
15	0·187	0·124	0·093	0·062
20	0·140	0·093	0·070	0·047
30	0·093	0·062	0·047	0·031

and assuming for $\varepsilon_r/\varepsilon_{0.2}$ the values computed in Table 5 according to the material and to the slenderness.

It should be noted that the relationship

$$\frac{\kappa_r}{\kappa_{0.2}} = \frac{\varepsilon_r}{\varepsilon_{0.2}} \tag{39}$$

is only valid in the case of profiles symmetrical with respect to the neutral axis. In the case of asymmetrical profiles the coefficient can be obtained by the relationship between the moment M and the maximum deformation ε_{max} of the highly stressed fibre. In this case eqn (38) becomes:

$$\frac{M}{M_{0.2}} = \frac{\varepsilon_{max}}{\varepsilon_{0.2}} - \frac{\varepsilon_r}{\varepsilon_{0.2}} \tag{40}$$

When this approach is used, the ψ value represents the intersection point between the $M-\kappa$ curve (in the symmetrical case) and the $M-\varepsilon_{max}$ curve (in the asymmetrical case) and the line represented by eqn (38) or eqn (40) respectively.

3.2.3 Fully plastic moment

(a) Extruded members. The moment–curvature relationship in elastic/perfectly plastic materials, such as steel, shows an asymptotic behaviour. The asymptote represents the fully plastic moment for increasing values of curvature.

In the case of elastic-hardening materials, such as aluminium alloys, this relationship is continuously increasing with a slope which is proportional to the degree of hardening of the alloy (Fig. 12(d)).

The value of the ultimate moment M_u, which gives the highest capacity of the cross-section, has to be related to the capacity of the material to deform without danger of early brittle failure. Therefore, the curvature has to be identified as the controlling ductility parameter by setting a limiting value of κ_{lim} for it. The ultimate moment is therefore:

$$M_u = M(\kappa_{lim}) \tag{41}$$

This relationship is not in a closed form; different κ_{lim} values can be taken for each alloy, and also the ductility requirements of each particular structure can be taken account of.

A conservative assumption is

$$\kappa_{lim} = 5-10\kappa_{0.2} \tag{42}$$

because the most unfavourable failure conditions for the less ductile materials can fall within this range. In steel structures, the fully plastic moment can be assumed as

$$M_{pl} = \alpha_p M_e \tag{43}$$

where

$$\alpha_p = \frac{Z}{W} \tag{44}$$

and is the ratio between the plastic and the elastic modulus of the cross-section (called the shape factor). Analogously, the plastic moment and the ultimate moment respectively in aluminium alloys can be related to the limit elastic moment by the relation

$$M_{pl} = \alpha_p M_{0.2} \tag{45}$$

$$M_u = \alpha M_{0.2} \tag{46}$$

where α is the generalized shape factor. α depends upon the cross-sectional geometry, the σ–ε law of the material (if the Ramberg–Osgood law is used the exponent n and $f_{0.2}$ are the main parameters) and upon the assumed κ_{lim} value.

The two shape factors, geometrical α_p and generalized α, with $\alpha > \alpha_p$, can be related through nomograms. On the horizontal axis are the α_p values which identify the geometry of the cross-section, and on the vertical axis are the values of α defined by eqn (46) as ratios between the ultimate moment M_u (computed for a given limit curvature κ_{lim}) and the elastic limit moment $M_{0.2}$. Each line characterized by a value of n (exponent of the Ramberg–Osgood law) and of κ_{lim} represents the points obtained by numerical simulation of the moment–curvature diagram and allows the actual shape factor α to be related to the conventional one α_p.

In the case of different aluminium alloys having a conventional elastic limit $f_{0.2}$ of 100, 200 and 300 N/mm^2 and the exponent n equal to 10, 20, 30, Figs 13(a,b,c) give the α_p–α relationships corresponding to the two limiting values of κ_{lim} proposed in eqn (42).

It should be noted that the line corresponding to $n = \infty$ starts at the origin and represents all the points where $\alpha_p = \alpha$ which is valid for elastic/perfectly plastic materials such as mild steels. As n decreases, that is as the hardening behaviour of the materials increases, the line rises and gives $\alpha > \alpha_p$, which shows the greater resistance in the inelastic range of aluminium alloys with respect to steel.

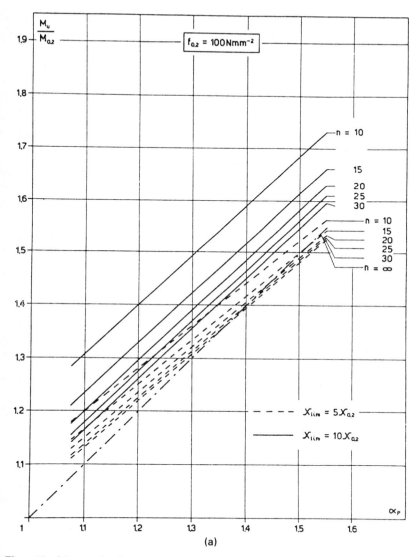

Fig. 13. Moment shape factor–maximum fibre strain–behaviour. (a) $f_{0.2} = 100 \text{ N/mm}^2$.

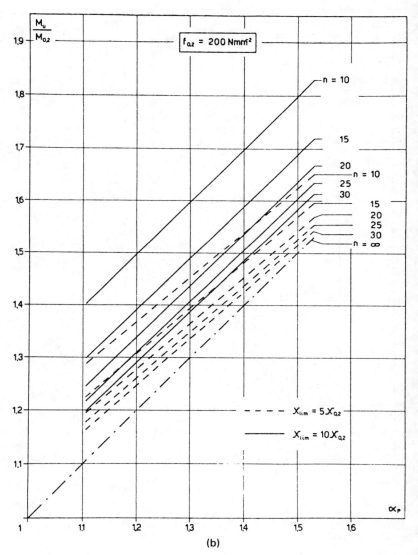

Fig. 13—*contd.* (b) $f_{0.2} = 200$ N/mm^2

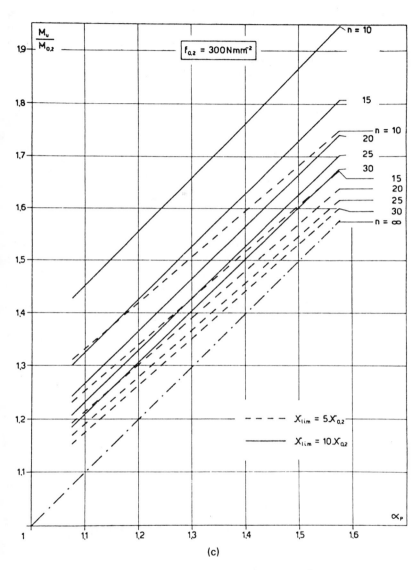

Fig. 13—*contd.* (c) $f_{0.2} = 300$ N/mm^2.

As an example, consider two double T sections with $\alpha_p = 1\cdot2$, one made of mild steel and the other of aluminium alloy. For steel, since $n = \infty$ for whatever type of mild steel, $\alpha = \alpha_p = 1\cdot2$. For aluminium alloy, the shape factor depends upon the type of alloy. In the case on non-heated-treated alloys, such as a 5*** series alloy with $f_{0\cdot2} = 160\,\text{N/mm}^2$, $n = 10$, the following is obtained from Fig. 13(a):

$$\alpha = 1\cdot28 \quad \text{for} \quad \kappa_{\lim} = 5\kappa_{0\cdot2}$$

$$\alpha = 1\cdot40 \quad \text{for} \quad \kappa_{\lim} = 10\kappa_{0\cdot2}$$

In the case of a heat-treated alloy, such as a 7*** series alloy with $f_{0\cdot2} = 300\,\text{N/mm}^2$, $n = 30$, the following is obtained from Fig. 13(c):

$$\alpha = 1\cdot26 \quad \text{for} \quad \kappa_{\lim} = 5\kappa_{0\cdot2}$$

$$\alpha = 1\cdot30 \quad \text{for} \quad \kappa_{\lim} = 10\kappa_{0\cdot2}$$

If the other variables are unchanged, the α/α_p ratio indicates the increase in strength of the conventional ultimate moment for a hardening material with respect to that of an elastic/perfectly plastic material.

A comparison between different definitions of M_u is given in Ref. 18, where a new formulation in a closed form is proposed.

A simplified method[19] to evaluate M_u of eqn (41) is based on the following formulae

$$M_u = (\alpha_p + \Delta\alpha)M_{0\cdot2} \qquad (47)$$

where

$$\Delta\alpha = \frac{\sigma(\varepsilon_{\lim}) - f_{0\cdot2}}{f_{0\cdot2}} \qquad (48)$$

where

α_p is given by eqn (4),

$M_{0\cdot2}$ is given by eqn (26).

The value of $\sigma(\varepsilon_{\lim})$ can be obtained by the Ramberg–Osgood law (eqn (12)) for a given strain $\varepsilon_{\lim}$ such as

$$\varepsilon_{\lim} = 5\text{–}10\varepsilon_{0\cdot2}$$

The comparison between the exact method[1] and the simplified formula (47) is given in Table 6. An additional comparison between both simplified methods[18,19] is shown in Table 7, where in the lines

Table 6. Comparison Between Exact Method and Simplified Method

n	$f_{0.2}$ (N/mm²)	σ_{u5} (N/mm²)	σ_{u10} (N/mm²)	$\sigma_p = 1.138$				$\sigma_p = 1.272$				$\alpha_p = 1.546$			
				α_5		α_{10}		α_5		α_{10}		α_5		α_{10}	
				1	19	1	19	1	19	1	19	1	19	1	19
11	110	121.77	131.15	1.232	1.245	1.333	1.330	1.347	1.379	1.464	1.464	1.551	1.653	1.703	1.738
14.5	145	159.65	168.92	1.231	1.239	1.307	1.303	1.350	1.373	1.442	1.437	1.562	1.647	1.685	1.711
20	200	217.96	227.07	1.222	1.228	1.278	1.273	1.346	1.362	1.415	1.407	1.565	1.636	1.660	1.681
21.5	215	233.70	242.77	1.221	1.225	1.272	1.267	1.344	1.359	1.409	1.401	1.564	1.633	1.654	1.675
23.5	235	254.61	263.63	1.217	1.221	1.265	1.260	1.342	1.355	1.402	1.394	1.563	1.629	1.648	1.668
24.5	245	265.03	274.03	1.216	1.220	1.262	1.256	1.341	1.354	1.398	1.390	1.562	1.628	1.644	1.664
26.5	265	285.82	294.79	1.213	1.217	1.256	1.250	1.339	1.351	1.392	1.384	1.561	1.625	1.638	1.658
28	280	301.38	310.31	1.210	1.214	1.250	1.246	1.337	1.348	1.388	1.380	1.559	1.622	1.634	1.654
Greatest deviation (%)				1.1		0.5		2.4		0.6		6.5		2.0	

Table 7. Additional Comparison Between Exact Method and Simplified Method

		Section 1 (50×80)	Section 2 (60×60)	Section 3 (80×50)	Section 4 (T-section)	Section 5 (L-section 40×80)
1	n	9.30	9.30	9.30	5.71	40.12
2	$f_{0.2}$	171.18	171.18	171.18	97.12	210.92
3	$M_{0.2}$	2 662	2 686	3 539	1 488	833
4	M_{pl}	3 123	3 225	4 378	1 709	1 501
5	ε_5	0.012 2	0.012 2	0.012 2	0.006 9	0.015 1
6	σ_{u5}	202	202	202	115	220.55
7	M_{u5}	3 598 3 602	3 616 3 708	4 959 5 015	2 023 1 983	1 474 1 539
8	ε_{10}	0.024 5	0.024 5	0.024 5	0.013 9	0.030 1
9	σ_{u10}	220.8	220.8	220.8	133	225.05
10	M_{u10}	3 958 3 895	3 926 4 004	5 479 5 404	2 364 2 259	1 541 1 557
11	ε_{max}	0.084	0.084	0.084	0.119 0	0.074 0
12	σ_{umax}	254.6	254.6	254.6	197.8	230.5
13	M_{umax}	4 571 4 420	4 660 4 534	6 339 6 103	3 252	1 617 1 578

14	n	26.56	26.56	26.56	26.56	16.81	39.51	49.33
15	$f_{0.2}$	171.18	171.18	171.18	171.18	97.12	225.63	210.92
16	$M_{0.2}$	2 662	2 686	3 539	1 488	4 012	833
17	M_{pl}	3 123	3 225	4 378	1 709	5 084	1 501
18	ε_5	0.0122	0.0122	0.0122	0.0069	0.0161	0.0151
19	σ_{u5}	181.5	181.5	181.5	103.1	236.45	218.7
20	M_{u5}	3 209 3 283	3 314 3 387	4 484 4 591	1 758 1 801	5 165 5 276	1 533 1 532
21	ε_{10}	0.0245	0.0245	0.0245	0.0139	0.0139	0.0301
22	σ_{u10}	187.3	187.3	187.3	108.25	241.4	222.35
23	M_{u10}	3 328 3 374	3 444 3 478	4 672 4 711	1 852 1 879	5 337 5 364	1 521 1 546
24	ε_{max}	0.084	0.084	0.084	0.1190	0.0553	0.0740
25	σ_{umax}	196.8	196.8	196.8	123.75	245	226.74
26	M_{umax}	3 499 3 521	— 3 627	4 916 4 908	2 133 2 117	5 420 5 428	1 563 1 563

Remarks: $\varepsilon_5 = 5\varepsilon_{0.2}$
$\sigma_{u5} = \sigma(\varepsilon_{lim} = 5\varepsilon_{0.2})$
M_{umax} corresponds to $\varepsilon_{lim} = 0.5\varepsilon_t$, ε_t being the collapse strain (1, p. 274).

M_{u5}, M_{u10}, M_{umax} the left values come from Ref. 18 and the right from Ref. 19. In both cases the agreement is very satisfactory.

A comparison of the full plastic range for the cross-sections T and C without welds (see Fig. 5(c)) is given in Fig. 14(a,b). The upper curve is valid for base material, the lower curve for a cross-section consisting completely of weld material. The intermediate curve will be discussed below.

(b) Welded members. The fully plastic moment M_u of welded members can be evaluated similarly to eqn (41) provided that the effects of the heat-affected zones are accounted for and attention is paid to the limitation of strain ε_{lim} due to the reduced ductility of the weld and the heat-affected material (see below).

In the case of members built up by longitudinal welds a simplified formula corresponding to eqn (47) can be adopted

$$M_{u,red} = \alpha_p W_{red} f_{0\cdot 2} + \Delta \alpha W f_{0\cdot 2} + \frac{\beta \Delta \alpha_w - \Delta \alpha}{1 - \beta}(W - W_{red}) f_{0\cdot 2} \quad (49)$$

α_p = shape factor, given by eqn (19),
W_{red} = reduced section modulus, given by eqn (33),
W = section modulus of the full section,
β = reduction factor, see Section 2.2,
$\Delta \alpha$ is evaluated by means of eqn (48) on the basis of the Ramberg–Osgood law (eqn (12)) for the base material at a given strain ε_{lim},
$\Delta \alpha_w$ is evaluated by means of eqn (48) on the basis of the Ramberg–Osgood law (eqn (12)) for the weld material at the given strain ε_{lim} and replacing $f_{0\cdot 2}$ by $f_{0\cdot 2,red}$.

The fixation of the value ε_{lim} must consider the fact that the weld material as well as the material of the heat-affected zone has a reduced ductility compared with the base material (see Section 3.4.3).

A comparison between the results of this formula and the exact solution (intermediate curve) coming from a computer simulation is given in Fig. 14(a,b) for the cross-sections T and C with longitudinal welds (see Fig. 5(c)). The agreement is very satisfactory for an outer fibre strain of $\varepsilon \geq 2\varepsilon_{0\cdot 2}$.

In the case of transverse welds the calculation methods for extruded members (see above) are still valid provided that the $f_{0\cdot 2}$ value is replaced by $f_{0\cdot 2,red}$ (see Section 2.2), $M_{0\cdot 2}$ by $M_{0\cdot 2,w} = W f_{0\cdot 2,red}$ and M_{pl}

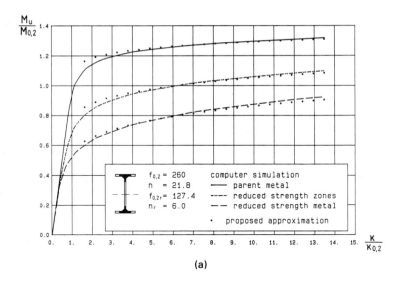

(a)

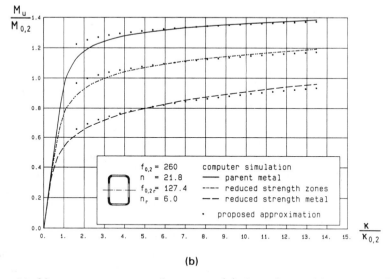

(b)

Fig. 14. Moment–curvature–diagrams. (a) I-section with welds. (b) Hollow section with welds.

by $M_{pl,w} = \alpha_p W f_{0.2,\text{red}}$ respectively and the ultimate moment capacity becomes $M_{u,w}$.

3.3 Plastic behaviour of statically undetermined girders

3.3.1 Extruded members

A step by step analysis has been applied to several structural layouts of beams with three different cross-sections, covering the range of possible shape factors from the extreme case of two concentrated masses to that of the rectangular section.

Three materials with different hardening behaviours have been examined:

—high hardening ($n = 8$),
—medium hardening ($n = 16$) and
—low hardening ($n = 32$).

The structural layouts considered were mainly symmetrical and asymmetrical continuous beams with two or three spans. Different values of spans have been considered in order to provide different maximum-moment/minimum-moment ratios and hence different ductility requirements.

Figure 15(a,b,c) show the behaviour of the asymmetrical schemes. The cases under consideration are:

(a) two equal spans,
(b) two unequal spans,
(c) one span completely restrained at one end and simply supported at the other end.

If we indicate with L_1 and L half the lengths of the spans, cases (a), (b) and (c) correspond to the ratios $L_1/L = 1 \cdot 0$, $0 \cdot 5$, $0 \cdot 0$. The continuous lines represent the behaviour of the simulated curve. The dashed lines represent the diagrams obtained for the same beams, with the assumption of elastic/perfectly plastic material, using the plastic hinge method, which assumes for the limiting moment the fully plastic moment with $f_y = f_{0.2}$. The discontinuities of these diagrams correspond to the occurrence of plastic hinges.

The loads and displacements are normalized with respect to their corresponding elastic limit values, which are reached when the $f_{0.2}$ stress is attained in the highly stressed section. In order to make more significant comparisons, the curves which cross each $F-v$ curve at the

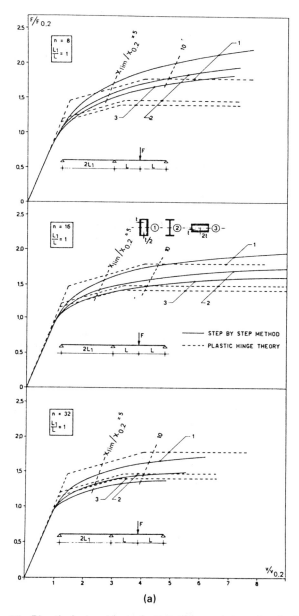

Fig. 15. Plastic behaviour of statically undetermined girders.

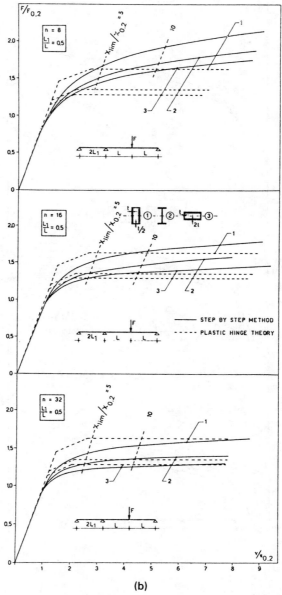

(b)

Fig. 15—*contd.*

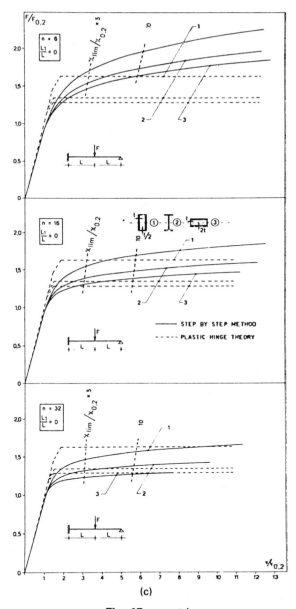

(c)

Fig. 15—*contd.*

points in which the highly stressed sections reach a limiting dimensionless curvature equal to $\kappa/\kappa_{0.2} = 5$ and 10 have also been shown.

This second limit is particularly significant because many structural aluminium alloys exhibit a ductility which just allows this curvature. In these cases this limit represents a threshold beyond which the material is no longer reliable. The analysis of these figures suggests two considerations with respect to the effects of the $\sigma-\varepsilon$ law and the shape of the cross-section on the moment–curvature relationship. With respect to the material it should be noted that the strong and medium hardening materials ($n = 8$, 16) reach greater values of resistance than predicted by plastic hinge theory at small curvatures. The same is not always true in the case of materials which show weak hardening ($n = 32$). On the other hand the cross-sections with large shape factors (cross-section 1) require a large plastic adaptation in order to take complete advantage of the strength of the section itself. Therefore, these curves cross the elastic/perfectly plastic curves at larger values of curvature.

With respect to the structural scheme influence on the redistribution process it should be noted that the higher the difference between the maximum and minimum elastic moments, and therefore the greater the need for redistribution of the moments (maximum in $L_1/L = 1$ and minimum in $L = 0$), the higher the required ductility of the material in order to reach limiting load values greater than those obtainable with the plastic hinge method (Fig. 15(a,b,c)).

The same behaviour is observed for all the aspects examined ($\sigma-\varepsilon$ law, shape of the cross-section, influence of the structural scheme: Fig. 16(a,b,c)), in symmetrical layouts such as a completely restrained beam, a beam with three different spans and a beam with three equal spans. Two cross-sections are considered; the intermediate shape factor is excluded.

These examples confirm the previous considerations, and in particular the influence of the cross-section, which for the same loading condition requires a higher ductility in the sections with large shape factors, especially in the case of materials showing little hardening. The influence of the loading condition, in contrast, seems to be less even when varying the ratio between the maximum and the minimum moment from 1·19 to 1·57.

The previous analysis shows that the factors which affect the plastic behaviour of continuous aluminium alloy beams can be classified in two different ways: with respect to the behaviour of the cross-section

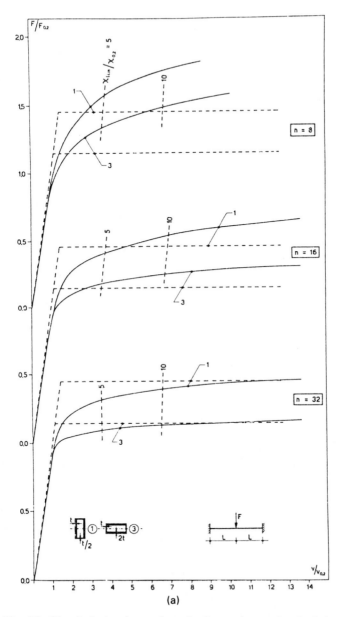

Fig. 16. Plastic behaviour of statically undetermined girders.

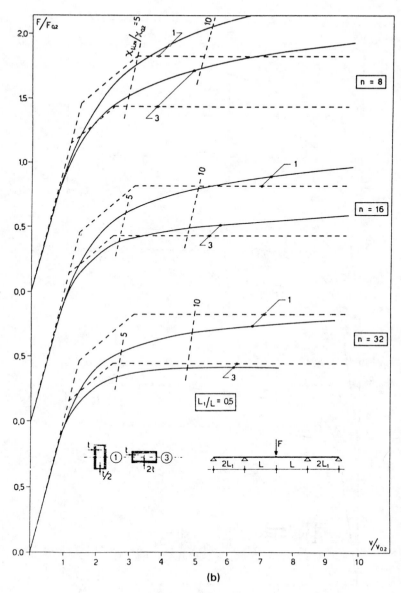

Fig. 16—*contd.*

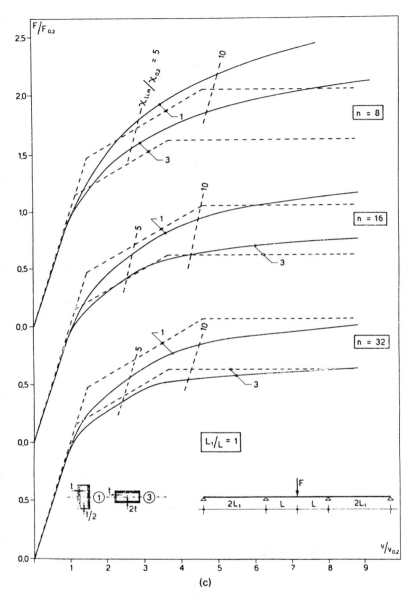

Fig. 16—contd.

and with respect to the influence of the structural layout and loading condition.

It should be noted that the effect of the cross-section is more significant, and it is therefore the first to be taken into account when attempting to apply the plastic hinge method to aluminium alloy structures.

As both the shape of the cross-section and the $\sigma-\varepsilon$ law affect the behaviour of the cross-section, it is useful to associate hardening and ductility parameters with all of the materials. This significant representation can be obtained by assigning to each alloy, through its characteristic parameters, a point on a diagram in which the axes represent hardening (n exponent of the $\sigma-\varepsilon$ law) and brittleness, the inverse of ductility.

Whereas the exponent n is given by a generalized expression which is accepted and dependent upon known parameters, such as $f_{0.2}$ and f_t, it is more difficult to define ductility in a precise way because it depends upon the uniform deformation ε_u, which is not usually known from the standard tensile test. The parameter for which data are usually known is the ultimate percentage elongation ε_t, which is measured after necking occurs. The value of ε_u is not strictly related to the ε_t value and is often 30% smaller.

If a conservative value is assumed of

$$\varepsilon_u \cong 0.5\varepsilon_t \tag{50}$$

then the parameter which is significant to the ductility of the material can be assumed:

$$\Delta = \frac{\varepsilon_u}{\varepsilon_{0.2}} = \frac{\varepsilon_t E}{2f_{0.2}} \tag{51}$$

which numerically expresses the extension of the inelastic behaviour of the material.

Under this hypothesis the alloys considered by the ECCS committee for Aluminium Alloy Structures have been drawn in Fig. 17(a) with their numerical designation. In the same figure it is also possible to represent, for a given alloy, a number of points which correspond to the ductility requirements for different structural layouts, loading conditions and cross-section shapes in order to obtain an ultimate load at least equal to that given by the plastic hinge method.

This analysis leads to three zones being defined in the figure. In the first region the materials are characterized by sufficient ductility and

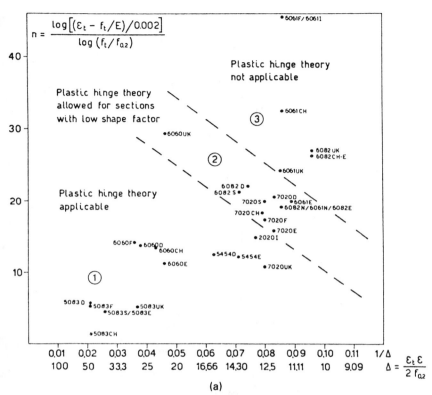

Fig. 17. Elastic and inelastic behaviour of different materials.

hardening to apply the plastic hinge method in all cases. In the second region, characterized by less ductility and hardening, the applicability of the method is limited to those sections with a small shape factor which do not require high plastic adaptation. In the third region, in contrast, the plastic hinge method usually provides limiting values greater than those allowable, and it is unsafe in these cases to use the plastic hinge method. This approach, though qualitative, can provide useful indications of the ductility requirements of different alloys.

The need for redistributing the bending moment beyond the elastic limit depends upon the structural layout (structure geometry and loading condition), whereas the possibility of doing so depends upon the ductility of the material.

A parameter which can characterize at the same time the redistribu-

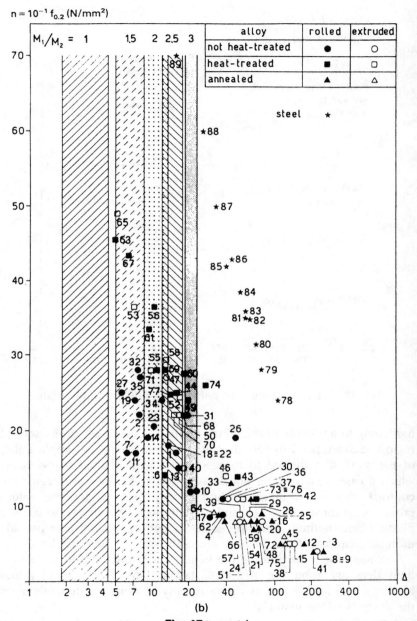

Fig. 17—contd.

tion requirements and the ductility requirements can be identified as the ratio between the maximum and minimum moments, M_1 and M_2, representative of all the structural layouts. From the analysis of different layouts, corresponding to M_1/M_2 values varying from 1 to 3 and for different values of the shape factor between the limits 1 and 1·5, it is evident that the representation in the n-plane (which relates the hardening and ductility values of the materials) identifies regions which can be considered independent of n.

On the basis of this consideration, if we assume $10n = f_{0.2}$, it is possible to provide an alternative, even if complementary, representation to that of Fig. 17(a). In this representation (Fig. 17(b)) the material ductility properties are superimposed on the redistribution requirements of the structural layout by means of the M_1/M_2 ratio. The symbols corresponding to several non-heat-treated (1–37) and heat treated alloys (37–77) are plotted in the same figure, together with stars representing the different mild and high strength steels (78–88).

The mechanical properties corresponding to these materials are given in Table 8. Each point, representative of a structural material, lies on a given redistribution range (M_1/M_2). This indicates that any structural layout, characterized by the M_1/M_2 value, can be evaluated at collapse by means of the plastic hinge method without danger of early brittle fracture. For example, alloy 67 can be used in all layouts with $M_1/M_2 \leq 1.5$, but it cannot be used when $M_1/M_2 > 1.5$.

Since in the commonly used structural layouts the value M_1/M_2 is always less than 3, it can be concluded that all the materials which fall in the unshaded region on the right of the figure can be used in plastic analysis without any particular limitation, whereas for the other materials the ductility requirements, which depend upon the structural layout (M_1/M_2), have to be taken into account.

In the range of activity of the ECCS Committee T2 (task group 2·2) further studies in aluminium alloy plastic design are now progressing, giving a general settlement of the theoretical assumptions[20] and making numerical and experimental applications[18,19,21,22] not only for extruded, but also for welded members.[16,19,23]

3.3.2 Welded members

In the case of a member constructed of longitudinal welds the above method can be applied provided that the ultimate bending moment of the cross-section is evaluated according to Section 3.2.3(b), welded

Table 8. Mechanical Properties of Metallic Materials

No.	Designation Formula	Designation Numerical designation	Stage	Thickness (mm)	f_1 (N/mm^2)	$f_{0.2}$ (N/mm^2)	ε_1 (%) As-rolled	ε_1 (%) Ex-truded
1	AlMn1·2Mg	3 004	H25	0·6–3	220	170	6	
2			H60		260	215	4	
3	AlMn1·2Cu	3 003	R	All	100	35	25	
4			H15	0·8–6	120	85	9	
5			H30	0·8–6	135	120	7	
6			H45	0·8–6	155	135	5	
7			H60	0·8–4	185	165	3	
8			Hp	—	100	40		25
9	AlMg0·8	5 005	R	All	100	40	25	
10			H25	0·8–6	135	120	8	
11			H60	0·8–4	185	165	4	
12	AlMg1·5		R	All	135	50	25	
13			H25	0·8–6	165	145	7	
14			H60	0·8–4	195	185	5	
15			Hp	—	135	50		20
16	AlMg2·5	5 052	R	All	155	80	22	
17			HL	8–75	165	85	7	
18			H20	0·8–6	215	175	7	
19			H40	0·8–4	255	235	5	
20			Hp	—	155	80		18
21	AlMg2·7Mn	5 454	R	All	205	80	16	
22			H20	0·8–6	245	175	7	
23			H40	0·8–4	265	200	6	
24			Hp	—	80			12
25	AlMg3·5		R	All	205	90	20	
26			H20	0·8–6	245	185	12	
27			H35	0·8–4	285	245	4	
28			Hp	—	215	90		16
29	AlMg4·4	5 086	R	All	245	110	20	
30			HL	8–75	255	110	12	
31			H15	0·8–6	285	215	12	
32			H30	0·8–4	325	275	6	
33	AlMg4·5	5 083	R	All	275	130	16	
34			H10	0·8–6	305	235	8	
35			H20	0·8–4	345	265	6	
36			Hp	—	265	110		12
37	AlMg5	5 056A	Hp	—	245	110		16
38	AlMgSi	6 060	TaN	≤3	135	50		20
39			TaA14	≤3	155	90		14
40			TaA16	≤3	195	145		8
41	AlSi1MgMn	6 082	R	All	90	40	25	
42			TN	0·8–4	205	110	22	
43			TA14	0·8–4	235	135	20	
44			TA16	0·8–4	295	245	11	
45			R	—	110	60		20
46			TA14	—	235	135		16

Table 8 —contd.

No.	Designation		Stage	Thickness (mm)	f_1 (N/mm^2)	$f_{0.2}$ (N/mm^2)	ε_1 (%)	
	Formula	Numerical designation					As-rolled	Ex-truded
47			TA16	—	315	265		10
48	AlCu4MgMn	2017 A	R	5–20	165	70	14	
49			THN	3·5–20	380	235	13	
50			TN	≤38	375	215		10
51	AlCu4·4SiMg	2 014	R	All	185	80		12
52			TN	All	345	240		12
53			TA	9–20	410	370		7
54	AlCu4·5MgMn	2 024	R	≤20	175	80	14	
55			THN	0·5–20	425	275	8	
56			TH06N	0·5–13	475	360	10	
57			R	All	195	80		11
58			TN	≤38	390	290		10
59	AlCu4·5MgMn		R	0·5–20	165	70	14	
60			THN	0·5–7	405	270	14	
61			TH06N	0·5–13	425	330	9	
62	AlZn4·8MgCu	7 075	MR	≤20	185	90	9	
63			TA	0·5–20	520	445	6	
64			R	All	185	90		9
65			TA	≤38	540	480		7
66	AlZn5·8MgCu		R	0·5–20	185	80	9	
67			TA	1–13	495	425	8	
68	AlZn4·5Mg	7 020	TaN	0·5–12	315	215	12	
69			TaA	0·5–12	355	275	10	
70			TaN	All	315	215		10
71			TaA	All	355	275		8
72	AlMg1SiCu	6 061	R	All	100	50	16	
73			TN	0·8–6	205	110	15	
74			TA16	0·8–6	295	245	10	
75			R	—	110	60		16
76			TN	—	175	110		16
77			TA16	—	265	235		9
78	Fe360			≤16	360	235	26	28
79	Fe430			≤16	430	275	23	24
80	Ex-TEN 45				445	310	21	
81	Ex-TEN 50				480	345	20	
82	COR-TEN A				480	345	22	
83	FE510			≤16	510	355	21	22
84	Ex-TEN 55				515	375	19	
85	COR-TEN C				550	410	20	
86	S420			<16	540	420	19	
87	S490			<16	570	490	16	
88	S590			<16	640	590	15	
89	S690			<16	780	690	14	

members, and attention is paid to the limitation of strain $\varepsilon_{\lim}$ due to the reduced ductility of the weld and the heat-affected material.

For members built up with cross welds, both the sequence of forming and the position of plastic hinges may be different from those of the previous cases for extruded members, due to the fact that the capacity of the welded section is considerably lower than that consisting of unaffected material.

In order to analyse the behaviour of beams with cross welds, it is useful to relate the bending moment $M(x)$ from the elastic calculation to the actual capacity of the cross-section at each position over the whole length of the girder, i.e. also at the welded cross-section. From this, the following relationships result

$$m_{\text{pl}}(x) = \frac{M(x)}{M_{\text{pl}}} \tag{52}$$

$$m_{\text{pl,w}}(x_w) = \frac{M(x_w)}{M_{\text{pl,w}}} \text{ at weld position} \tag{53}$$

If $m(x_w) > m(x)$, one can be sure that the first hinge will occur in the cross welded section, before it forms in any other section. With increasing loads a plastic hinge mechanism occurs as shown in Fig. 18(a), different from that usually assumed. Since the strain in the most highly strained fibre of the cross weld must be limited to a certain value $\varepsilon_{\lim}$ to prevent rupture, the ultimate reduced moment $M_{u,w}$ can be evaluated according to Section 3.2.3. From this a limited rotational capacity in this hinge results, consequently the full hinge mechanism cannot occur. The theory is based on the fact that at this stage the moment δM_u is present at the position where the second plastic hinge would occur. Using a procedure described in Ref. 19 the value δ can be derived from the relationship

$$\delta = \frac{\min m}{\max m} \geq 0 \tag{54}$$

where

$$\max m = \max \begin{cases} |m_{\text{pl,w}}| \\ |m_{\text{pl}}| \end{cases}$$

$$\min m = \min \begin{cases} |m_{\text{pl,w}}| \\ |m_{\text{pl}}| \end{cases}$$

The practical application of this method for cross welded beam is shown in the following two examples.

- hinge in the unaffected beam
- hinge in the weld

1. weld in the range of positive moments

$m_{plw} > m_{pl12}$

$m_{plw} < m_{pl12}$

2. weld in the range of negative moments

$m_{plw} > m_{pl2}$

$m_{plw} < m_{pl2}$

(a)

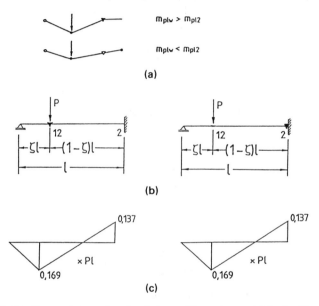

Fig. 18. Statically undetermined girder with cross welds in the plastic range. (a) System and loading, position of welds and plastic hinge. (b) M-distribution for weld and load at point 12. (c) M-distribution for weld at the restraint.

Example 1 (Fig. 18(b))
Geometrical data:
Cross-section: I-shape (see Fig. 21(a), below),
Length of the girder: $l = 3 \cdot 00$ m,
Position of the cross weld: $\zeta = 0 \cdot 3$

Material data:
$f_{0\cdot2} = 314\cdot1\,\text{N/mm}^2$
$f_{0\cdot2,\text{red}} = 147\cdot4\,\text{N/mm}^2$

Capacity data:
Extruded section:
$M_{0\cdot2} = 78\cdot12\,\text{kNm}$
$M_{\text{pl}} = 90\cdot32\,\text{kNm}$
$M_{\text{u}} = 100\cdot17\,\text{kNm}$

Cross welded section:
$M_{0\cdot2,\text{w}} = 36\cdot66\,\text{kNm}$
$M_{\text{pl,w}} = 42\cdot39\,\text{kNm}$
$M_{\text{u,w}} = 64\cdot18\,\text{kNm}$

Related moments:
Cross welded section:
$$m_{\text{pl,w}} = \frac{0\cdot169\,pl}{42\cdot39} = 3\cdot99 \times 10^{-3}\,pl$$

Extruded section:
$$m_{\text{pl,2}} = \frac{0\cdot137\,pl}{90\cdot32} = 1\cdot52 \times 10^{-3}\,pl$$

Since $m_{\text{pl,w}} > m_{\text{pl,2}}$, the first plastic hinge will form in the cross welded section at the position 12, and M_{12} can go up to $M_{\text{u,w}}$. With increasing load and the limitation of the mostly strained fibre in the weld to ε_{lim} the value M_2 at position 2 reaches the value $\delta_2 M_{\text{u}}$.

According to the theory the value δ_2 is approximately

$$\delta_2 = \frac{\min m}{\max m} = \frac{|m_{\text{pl,2}}|}{|m_{\text{pl,w}}|} = 0\cdot380$$

The ultimate load P_{u} can be evaluated by means of the principle of virtual work, giving

$$P_{\text{u}} = \frac{M_{\text{u,w}} + \zeta\delta_2 M_{\text{u}}}{\zeta(1-\zeta)l} = 120\cdot0\,\text{kN}$$

Using a more advanced method and a computer simulation, the exact ultimate load is $P_{\text{u}} = 141\cdot2\,\text{kN}$, while the elastic load when reaching $M_{0\cdot2,\text{w}}$ in 12 is $P_{\text{el}} = 75\cdot1\,\text{kN}$.

Example 2 (Fig. 18(c))
The geometrical, material and capacity data are the same as in Example 1 with the exception of the position of the cross weld, which is now placed on position 2.

The related moments are as follows

Cross welded section: $$m_{\text{pl,w}} = \frac{0\cdot137\,pl}{42\cdot39} = 3\cdot23 \times 10^{-3}\,pl$$

Extruded section: $m_{pl,12} = \dfrac{0.169 \, pl}{90.32} = 1.87 \times 10^{-3} \, pl$

Since $m_{pl,w} > m_{pl,12}$ the first plastic hinge will form in the cross welded section 2 and the load can be increased until the ultimate moment $M_{u,w}$ is reached in this position. In this situation the moment in section 12 is smaller than M_u for the extruded profile and according to the theory it is approximately $\delta_{12} M_u$ where

$$\delta_{12} = \frac{\min m}{\max m} = \frac{|m_{pl,12}|}{|m_{pl,w}|} = 0.579$$

By means of the principle of virtual work, P_u can be calculated from the following equation

$$P_u = \frac{\delta_{12} M_u + \zeta M_{uw}}{\zeta(1-\zeta)l} = 122.6 \text{ kN}$$

Using the conventional elastic theory the maximum load is $P = 89.2$ kN.

3.4 Flexural torsional buckling

3.4.1 Physical behaviour

The buckling behaviour of beams is usually referred to as flexural torsional buckling.

Consider a deep plate girder with two end moments. As the applied load increases, the vertical displacements increase and the deflection of the beam remains in the same plane. For a given value of the load, defined as the critical load, the beam suddenly buckles laterally and twists (Fig. 19).

In real structures the presence of geometrical and mechanical imperfections of the industrial bar (see Section 1) makes the process more gradual because there is no bifurcation of equilibrium. The deformations develop gradually with the increase of loads beyond the elastic range up to collapse conditions, and these identify the maximum load bearing capacity of the beam.

Flexural torsional buckling has been studied extensively both in the elastic and in the elastoplastic range, especially with respect to steel structures.

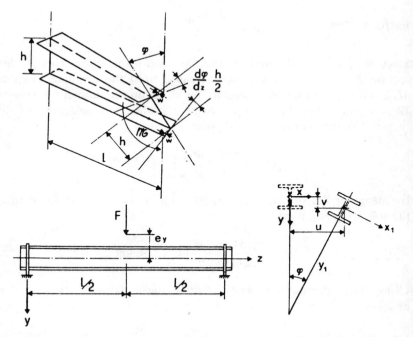

Fig. 19. Deformation of a cantilever beam with end load due to lateral torsional buckling.

3.4.2 *Experimental and theoretical results*

Elastic bifurcation theory leads, as is well known, to the following expression for the critical bending moment of double T sections:

$$M_{cr,D} = \psi_1 \frac{\pi}{l_{c,h}} \sqrt{EI_y GI_T} \sqrt{1 + \frac{\pi^2}{k^2}} \tag{55}$$

where

$$k = L_{c,h}\sqrt{GI_T/(EI_\omega)} \tag{56}$$

and

ψ_1 = a coefficient depending upon the load distribution and boundary conditions,
$L_{c,h}$ = effective length of the girder, which is represented by the distance between points of inflection,
EI_y = lateral bending rigidity,
GI_T = Saint Venant's torsional rigidity.
EI_ω = warping torsional rigidity.

Equation (55), which was first developed for steel structures, can also represent the behaviour of aluminium alloy beams.

Complex simulation programs were required in order to investigate the inelastic range including the presence of imperfections. Even though the aspects of the problem were analysed, no practical design rules were developed in this manner. A method which was useful for this purpose is that proposed by Lindner[24] for steel structures. The bending moment which causes buckling of a beam is given by:

$$M_D = M_{pl} \left[\frac{1}{1 + \bar{\lambda}_M^{2n}} \right]^{1/n} \quad (57)$$

where

M_{pl} = plastic moment,
n = 'system factor',

and

$$\bar{\lambda}_M = \sqrt{\frac{M_{pl}}{M_{cr,D}}} \quad (58)$$

is the bending slenderness ratio, where $M_{cr,D}$ is the critical elastic moment which can be calculated from eqn (55).

This approach can also be used in aluminium alloys provided that the appropriate value of n is used. In order to use eqn (57) for aluminium alloys, it was necessary to calibrate the n coefficient on the basis of the available experimental results, which were neither recent nor numerous. In the past, several tests were carried out by ALCOA in the USA, and in particular:

—Dumont and Hill undertook tests on rectangular rolled sections[25] and on extruded double T sections[26]
—Hill undertook tests on asymmetrical double T profiles,[27]
—Clark and Jombock undertook tests on double T 2014-T6 profiles with a linearly varying bending moment,[28]
—Clark and Rolf undertook tests on 2*** series rectangular sections, the alloy being heat-treated and non-heat-treated.[29]

More recently some tests have been carried out in Germany by Klöppel & Bärsch[30] on double T AlMgMn and AlZnMgl profiles.

The above test results are collected in Ref. 1 in the form of buckling curve diagrams (Fig. 20(a–f)). The n coefficient which seems to

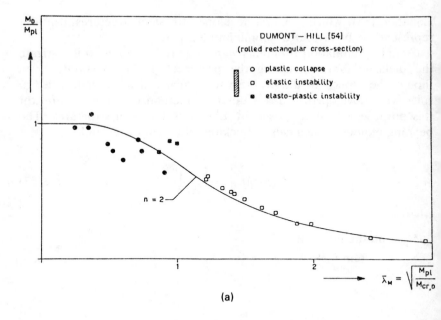

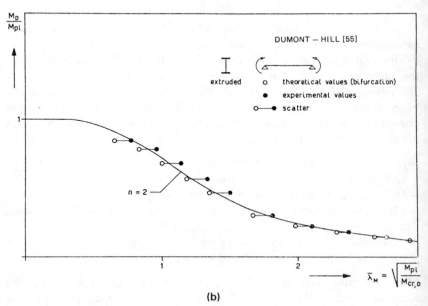

Fig. 20. Test results from bending tests. (a) and (b) Dumont–Hill.

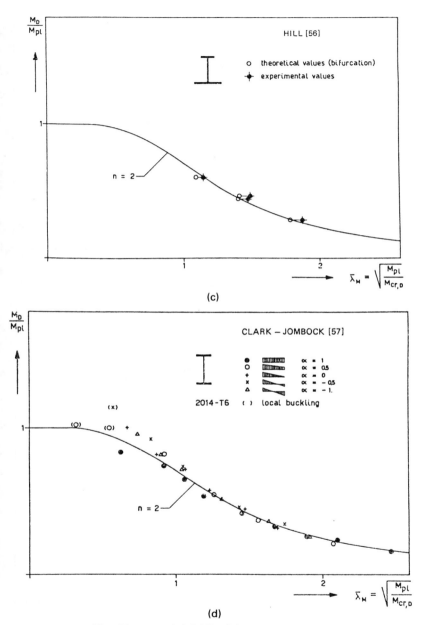

Fig. 20—*contd.* (c) Hill. (d) Clark–Jombock.

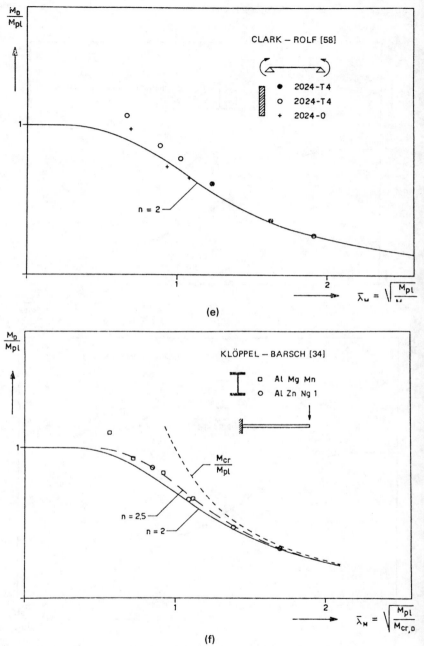

Fig. 20—*contd.* (e) Clark–Rolf. (f) Klöppel—Barsch.

interpret the experimental results conservatively is $n = 2$.[31] The ECCS committee decided to provisionally adopt this value in the first edition of the recommendations on aluminium alloy structures while waiting for further experimental results. The committee also decided to simplify M_{pl} computations assuming (as in steel structures)

$$M_{pl} = \alpha_p W f_{0 \cdot 2} \tag{59}$$

where

α_p = geometrical shape factor of the cross-section,
W = bending section modulus,
$f_{0 \cdot 2}$ = conventional elastic limit of the material.

The highest deviations of the formula from the experimental values are in the range $0 < \bar{\lambda}_M < 0 \cdot 75$ in which local buckling phenomena are of extreme importance since they cause local buckling of compressed parts.

Even though some experimental results fall below the analytical curve, this is not of major concern since actual beams usually have support conditions more restrained than those of the experimental tests. The proposed curve, with $n = 2$, seems appropriate to represent the averaged behaviour of the following cases:

—rectangular sections,
—double T (symmetrical and asymmetrical) sections (with $b_{max}/b_{min} \leq 1 \cdot 5$),
—rolled and extruded sections,
—heat-treated and non-heat-treated aluminium alloys,
—all variations of the bending moment diagram.

The last possibility is connected to the structure of eqn (57) because it indirectly takes account of loading and support conditions through the computation of $M_{cr,D}$, given by eqn (55).

3.4.3 Simulation results

The computer simulation of the load carrying behaviour of aluminium alloy members is, as also later shown, a very powerful aid. The recently published elaboration[16] allows the following parameters, to be taken into account:

—any open cross-section,
—non-linear stress–strain laws such as for aluminium,

—reduced non-linear stress–strain laws in any finite element, near the weld and in the heat-affected zone caused by the heat input,
—geometrically non-linear behaviour, i.e. second order theory,
—overall buckling,
—any end condition such as hinged supports, fully or partly restraints, free,
—any point load or uniformly distributed load parallel to the axis of the member and perpendicular to it,
—any support by discrete or continuous springs with linear elastic behaviour at the ends or within the length of the member,
—any imperfection such as initial deflections in both the principal axes, initial twist, end levers, residual stresses, unequal distribution of the E-modulus, etc.

Simulations have been carried out for an I-shaped cross-section Type A, which is shown in Fig. 21(a). To investigate the influence of an alteration in the depth/width ratio several other derivations of the cross-section type A have been treated; their dimensions are shown in Table 9.

The following initial geometrical imperfections have been introduced:

—an $L/1000$ deflection in the x-direction, constant over the length of the bar,
—an $L/1000$ deflection in the y-direction, constant over the length of the bar,
—an $L/(1000\,h)$ twist midspan, sinusoidally distributed so that the compression flange deflects in the same y-direction as before (unfavourable case).

Simulations have been carried out for the following materials and stress strain laws:

(a) base material for the whole bar,
   (i) with the point by point σ–ε data from the tests (see Section 1.4.2),
   (ii) with the Ramberg–Osgood law and the mechanical data for the material M1 (see Table 3),
   (iii) with the Ramberg–Osgood law and the mechanical data E and $f_{0.2}$ from ERAAS;
(b) weld material and HAZ material in and around the weld for a welded bar,

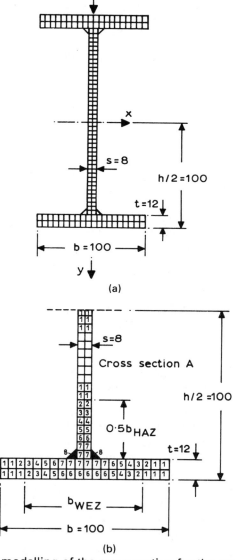

Fig. 21. Element-modelling of the cross section for the computer simulation of the buckling capacity. (a) Prototype for the element modelling. (b) Element modelling of an I-cross section with welds, the numbers in the heat affected zone around the weld represent elements with different stress–strain laws according to their distance to the weld.

Table 9. Element Modelling of the Cross-section for the Computer Simulation of the Buckling Capacity. Variation of the Dimensions of the I-Cross Section for the Simulation

Cross-section Type	Depth h (mm)	Width b (mm)	Flange thickness t (mm)	Web thickness s (mm)
A	200	100	12	8
B	500	200	30	16
C	500	200	16	10
D	200	200	30	16
E	200	200	12	8
F	500	500	24	14
G	200	100/50	12	8

(i) with the different point by point $\sigma-\varepsilon$ data from the tests for the elements according to their distances from the weld (see Fig. 21(b) for longitudinal fillet welds between the web and the flanges),

(ii) with the Ramberg–Osgood law and the mechanical data for the materials M1 up to M8, depending on the distance of the finite element to the weld (see Fig. 21(b)),

(iii) with two different Ramberg–Osgood laws for the base material and for the material of a 50 mm HAZ taking E, $f_{0.2}$ and $f_{0.2,\text{red}}$ from ERAAS,

(c) HAZ material M7 for the whole bar.

For the computer simulation the members with longitudinal welds as shown in Fig. 21(a) had to be divided lengthwise; in the actual case it was decided to have 20 sections of equal length. In the case of members with cross welds the spacing of the sections around the weld and the HAZ was such that the different $\sigma-\varepsilon$ laws such as M1–M8 could be taken into account in a very detailed way. Furthermore the cross-section has been split up into finite elements according to Fig. 21(b), and each element has been combined with its corresponding $\sigma-\varepsilon$ law depending on the distance to the weld. These data were taken from the tests as reported in Section 1.4.2 (cases b(i) or b(ii)).

The diagrams hereafter show computer simulated buckling curves for the flexural torsional buckling of the following members under bending moments:

—members consisting of pure base material,

—members having a cross weld at $x = 0.5 L$ or at $x_1 = 0.25 L$ and $x_2 = 0.75 L$ or at $x_1 = 0.125 L$ and $x_2 = 0.875 L$,
—members having longitudinal fillet welds between the web and the flanges,
—members consisting of pure HAZ material.

On the horizontal axis we read

$$\bar{\lambda}_M = \sqrt{M_{pl}/M_{cr,D}} \qquad \text{(see 58)}$$

on the vertical axis

$$\bar{M} = M_u/M_{pl}$$

where

$M_{cr,D}$ is from eqn (55) for a constant moment diagram, $\psi_1 = 1$,
M_{pl} is from eqn (45), usually evaluated with $f_{0.2}$ of the base material,
M_u is the carrying capacity.

In each diagram the corresponding buckling curve for the member consisting of pure base material is shown for comparison.

The first diagrams (Fig. 22(a–c)) allow us to answer the following questions:

1. What is the overall buckling capacity of an extruded beam under various load patterns?
2. What is the accurate drop down in capacity due to a cross weld midspan or in other places?
3. What is the influence of the moment gradient on the capacity?
4. Which advantage can be gained by not employing the simplest method of using pure HAZ material for the whole bar?

Answers:

Ans. 1. The buckling capacities for the three load patterns are quite equal.
Ans. 2. The drop down in the capacity of a girder with cross welds compared with that of a girder of pure base material is important for slenderness ratios up to 1·5. At very low $\bar{\lambda}_M$ values it reaches the value $f_{0.2/,red}/f_{0.2}$ for a constant moment distribution.

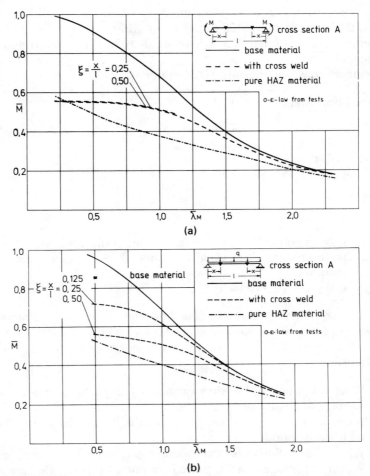

Fig. 22. Lateral buckling curves for beams with I-section consisting of base material or having cross welds at different positions or consisting purely of HAZ material. (a) Loaded by a constant moment diagram. (b) Loaded by uniformly distributed load.

Ans. 3. The moment gradient has a non-negligible influence on the capacity. If it is zero (Fig. 22(a)) the cross weld lowers the capacity equally, independent of its position. But in case of a parabolic (Fig. 22(b)) or linear (Fig. 22(c)) moment distribution the effect of the cross weld on the capacity decreases the nearer it is to the end of the girder.

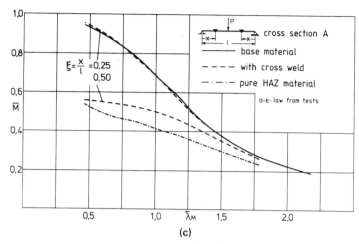

Fig. 22—*contd.* (c) Loaded by a single load midspan.

Ans. 4. The lowest curve in each diagram represents the buckling capacity of a girder consisting of pure HAZ material. It is very uneconomic for $\bar{\lambda}_M \leqslant 1\cdot 75$.

Another result should be mentioned here. The deflection of a girder with a cross weld in an area of substantial bending moment differs completely from that of a girder consisting of pure base material. The first has a sharp curvature concentrated at the weld position with very low curvature in the other parts while the latter has a sinusoidal-like deflection. At the sharp bend around the weld and in the HAZ, the strain in the outer fibre of the weakened part is very high, here greater than 7% (in the range of $0\cdot 8 \leqslant \bar{\lambda}_M \leqslant 1\cdot 0$) and far into the strain hardening range (see Table 10). The value of the maximum strain depends, of course, on the extent of the weakened zone. There is a need for limiting the actual strain of the most strained fibre in the weld to prevent local buckling of the outstands and rupture. With regard to this problem the following questions should be answered:

Ans. 5. How much does the extent of the weakened zone around the cross weld in direction of the length of the bar influence the carrying capacity?

Ans. 6. What decrease in load may be expected for the girders of type A and C of different depths, if ε_{lim} is fixed at a value of 3·5% in the most strained cross weld or HAZ fibre?

Table 10. Maximum Strain as a Percentage of the Outer Fibre in a Beam Under a Constant Moment Diagram

M	Base material $\sigma-\varepsilon$ law: M1	Pure HAZ material $\sigma-\varepsilon$ law: M7	Cross weld midspan $\sigma-\varepsilon$ law: M8	Longitudinal weld $\sigma-\varepsilon$ law: M1–M8
0·611	0·793	1·566	6·867	0·891
0·793	0·637	0·989	7·914	0·756
1·115	0·514	0·501	5·713	0·560
1·386	0·545	0·319	3·173	—
1·619	0·483	0·272	1·809	0·564

The answers may be given by Fig. 23 and the Fig. 24(a,b):

Ans. 5. Figure 23 shows the non-dimensional carrying capacity of the girder type B (very high depth) with a cross weld midspan. The codes which adopt a weakened zone of 50–60 mm length with a constant $f_{0\cdot 2,\text{red}}$ are on the safe side compared with exact simulation results.

Ans. 6. From Fig. 24(a,b) it can be seen that if $\varepsilon_{\text{lim}} = 3\cdot 5\%$ in the weld the decrease in load reaches as a maximum c. 10% in the range of $0\cdot 5 \leq \bar{\lambda}_M \leq 1\cdot 0$ (full lines). There is no collapse

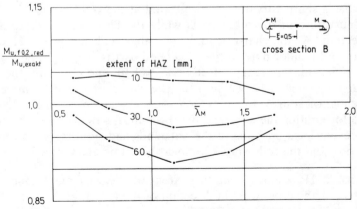

Fig. 23. Influence of the extent of the strength reduced zone (HAZ) between 10 and 60 mm on the lateral torsional buckling capacity of a girder with a cross weld midspan under a constant moment diagram.

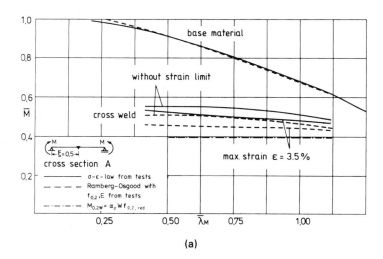

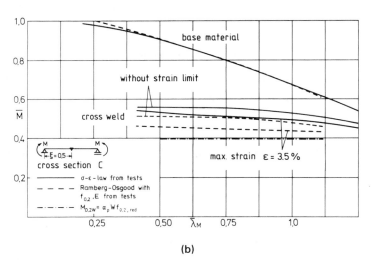

Fig. 24. Comparison between the lateral torsional buckling capacity of a girder with I-section with a cross weld midspan under a constant moment diagram and the carrying capacity of the same girder if the strain in the outer fibre of the HAZ is limited to 3·5%. (a) Cross section type A according to Table 9. (b) Cross section type C according to Table 9.

since due to $\varepsilon_{\lim} = 3.5\%$ the limit load is smaller than the buckling load.

Another two questions of interest are:

7. How much does the shape of the cross-section influence the overall buckling load?
8. What is the influence of monosymmetry in the cross-section?

This may be answered by Fig. 25(a,b) from which the following conclusions can be drawn:

Ans. 7. The buckling curves of girders with a type A (depth = 2 × width) or type B (depth = 2.5 × width) cross-section are very close together and represent the worst case among the investigated beams. The type D girder (depth = width) has a higher buckling curve for $0.5 < \bar{\lambda}_M < 1.25$ for base material and for $0.75 < \bar{\lambda}_M < 1.75$ for members with a cross weld midspan (Fig. 25(a)).

Ans. 8. The carrying capacity of a beam with the monosymmetrical cross-section type G and a weld midspan is represented by the same buckling curve as for the equivalent beam type A if the wider flange is in compression, it is about 10% lower if the smaller flange is in compression.

In all codes the heat affected zone is defined with a length of 50–60 mm and related to a yield strength of $f_{0.2,\text{red}}$ (see Section 2.2). This is on the safe side, but the question is:

9. What is the loss in the buckling load capacity for girders with cross welds in several positions under uniformly distributed load, if the actual $\sigma-\varepsilon$ laws M1–M8 (see Table 3) over the weakened zone are compared with a constant $f_{0.2,\text{red}}$ value and $10\, n = f_{0.2}$ for the base material and $10\, n_{\text{red}} = f_{0.2,\text{red}}$ for the weld and the HAZ material with the Ramberg–Osgood law?

The answer may be taken from the Fig. 26:

Ans. 9. The full lines, which are valid for the mechanical properties from the tests have already been shown in Fig. 22(b). The buckling curves for the ERAAS data (dashed lines) are lower between $0 \leq \bar{\lambda}_M \leq 1.25$ since $f_{0.2,\text{red,ERAAS}}$ is considerably lower than $f_{0.2,\text{red,tests}}$.

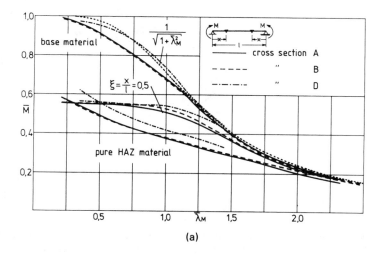

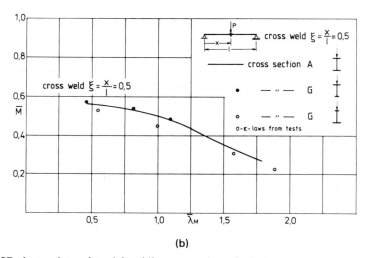

Fig. 25. Lateral torsional buckling capacity of girders with I-sections of different types (see Table 9) and cross welds midspan, comparison to equivalent girders consisting purely of base material or purely of HAZ material. (a) Double symmetrical cross sections type A, B, D. (b) Cross-sections type A (double symmetrical) and types G (monosymmetrical, top: wider flange in compression, bottom: smaller flange in compression).

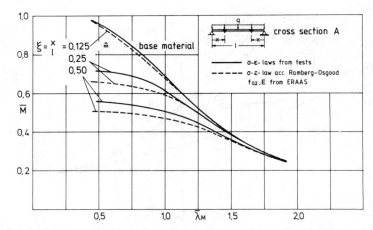

Fig. 26. Influence of the position of cross welds along the girder on the lateral torsion buckling load of a girder under a constant moment diagram.

The computer simulations have been extended to beams with longitudinal welds between the web and the flanges to show the influence on the overall buckling behaviour. It has been demonstrated in the literature,[11,12] that in members with longitudinal welds the residual stresses cannot be neglected. Therefore in the following simulations residual stresses of the pattern shown in Fig. 27 have been taken into

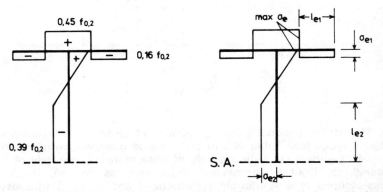

Fig. 27. Simplified residual stress distribution in a girder with I-cross section—type A with longitudinal welds between web and flanges.

account. The four questions to be answered are:

10. How much do longitudinal welds influence the buckling load?
11. What is the influence of the moment gradient?
12. What is the influence of the shape of the cross-section?
13. Are the assumptions of 50 mm for the extent of the HAZ and the $f_{0.2,\mathrm{red}}$ values in the codes sufficient and how different are the results from the accurate ones?

The solutions of these questions may be taken from Fig. 28(a,b,c).

Ans. 10. The loss of buckling capacity occurs especially in the range $0 < \bar{\lambda}_M \leq 1.25$, and at $\bar{\lambda}_M = 0$ it reaches the value A_{red}/A. For higher λ_M values the influence becomes zero, and in this range all curves approach the Euler buckling curve. It is not economic to replace all material by HAZ material.

Ans. 11. There is a little influence of the moment gradient on the buckling load. The welded girder with a concentrated load midspan reaches up to 10% more capacity, especially for $\bar{\lambda}_M < 1.0$, than the same girder under a constant bending moment.

Ans. 12. Figure 28(b) shows that there is only little influence due to the shape of the cross-section, and a difference is only noticeable for $\bar{\lambda}_M < 0.7$.

Ans. 13. Figure 28(c) presents the ratio of the ultimate bending moment $M_{u,\mathrm{red,ERAAS}}$ evaluated with mechanical data from ERAAS and the ultimate bending moment $M_{u,\mathrm{red,exact}}$ from computer simulations with the test data (see Section 1.4.2). To fix the extent of the HAZ equal to 50 mm is sufficient, sometimes 10% on the safe side for $0.5 < \bar{\lambda}_M < 1.0$; a value of 40 mm would just be an optimum.

3.5 Codification

3.5.1 Plastic design

In spite of the fact that much research has been developed to analyse the behaviour of aluminium alloy structures beyond the elastic limit this subject is not yet sufficiently investigated to give simple codified rules for practical application. For this reason plastic design is not yet included in modern codes for aluminium alloys.

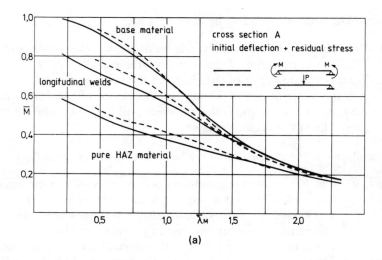

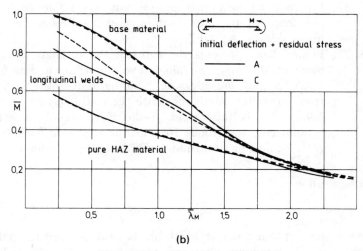

Fig. 28. Lateral torsional buckling capacity of girders with I-sections and longitudinal welds between web and flanges compared with purely base and purely HAZ material. (a) Influence of different moment diagrams. (b) Influence of different cross sections, type A and C.

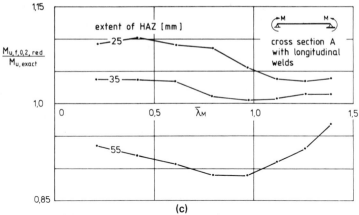

Fig. 28—*contd.* (c) Influence of the width of the heat affected zone around the weld (between 25 and 55 mm).

3.5.2 Buckling check

The ECCS method is based upon the application of eqn (57) with the exponent $n = 2$ (not to be confused with the value n in the Ramberg–Osgood law). The range of applicability of this formula is based on the following conditions:

—the cross-section has double symmetry (double T sections with equal flanges);
—loads act in the plane of the web;
—distortion of the cross-section and local buckling phenomena are not taken into account;
—supports do not allow lateral displacements and rotations outside the bending plane.

One has to check that:

$$M_{max} \leq \begin{cases} M_D \\ Wf_d \end{cases} \quad (60)$$

where

- M_{max} = maximum bending moment caused by design loads,
- M_D = ultimate flexural torsional buckling moment,
- W = section modulus of the cross-section,
- f_d = design strength of the material which is assumed equal to the conventional elastic limit $f_{0.2}$.

The M_D values are given by the relationship

Table 11. Non Dimensional M_D/M_{pl}-Values for Lateral Torsional Buckling in Relation to the Slenderness Ratio $\bar{\lambda}_M$

$\bar{\lambda}_M$	0	0·1	0·2	0·3	0·4	0·5	0·6	0·7	0·8	0·9
0	1	1	0·999 2	0·996 0	0·987 4	0·970 1	0·940 9	0·898 0	0·842 3	0·777 1
1	0·707 1	0·637 0	0·570 4	0·509 2	0·454 5	0·406 1	0·363 9	0·327 0	0·294 9	0·267 0
2	0·242 4	0·221 1	0·202 3	0·185 7	0·171 1	0·158 0	0·146 3	0·135 9	0·126 5	0·118 1
3	0·110 4	0·103 5	0·097 2	0·091 4	0·086 2	0·081 4	0·076 9	0·072 9	0·069 1	0·065 6
4	0·062 4	0·059 4	0·056 6	0·054 0	0·051 6	0·049 3	0·047 2	0·045 2	0·043 4	0·041 6
5	0·040 0	0·038 4	0·037 0	0·035 6	0·034 3	0·033 0	0·031 9	0·030 8	0·029 7	0·028 7

$$M_D = \frac{M_{pl}}{\sqrt{(1 + \bar{\lambda}_M^4)}} \tag{61}$$

where $\bar{\lambda}_M$ is the slenderness ratio given by eqn (58), and M_{pl} and $M_{cr,D}$ are given by eqns (59) and (55) respectively.

The values of the M_D/M_{pl} ratio are given in Table 11. When the value of $M_{cr,D}$ cannot be calculated exactly, we can use the following approximate method.

When a double T symmetrical or asymmetrical section deflects in the plane of the web, an analysis can be carried out by checking the buckling of the compressed flange assuming it to be isolated from the web. In fact it can be assumed that buckling is due to the compressed flange. The problem is then reduced to that of a bar under compression, and the method of Section 4.3 can be used.

One has to check that:

$$\sigma = \frac{N_f}{A_f} \leq f_d \bar{N} \tag{62}$$

where

- f_d = design strength of the material,
- A_f = area of the compression flange,
- $N_f = \dfrac{M_{eq}}{I_x} S_x$ is the axial force acting on the flange assumed independent from the web, under design loads,
- I_x = moment of inertia of the entire cross-section with respect to the axis of gravity x,
- S_x = static moment of the compression flange with respect to the x-axis,
- M_{eq} = reference equivalent moment.

The reference equivalent moment is computed from the maximum and the average bending moment in the portion of beam considered in the following way:

$M_{eq} = 1 \cdot 3 M_M$ in the case of simply supported or continuous beams, with the limitation $0 \cdot 75 \, M_{max} \leq M_{eq} \leq M_{max}$,

$M_{eq} = M_m$ in the case of cantilever beams, with the limitation $0 \cdot 5 \, M_{max} \leq M_{eq} \leq M_{max}$.

This approximate method, first used for steel structures, is physically immediate and allows the limitation of double symmetry to be overcome. It is included in the Italian code.

4 MEMBERS IN COMPRESSION

4.1 General

The load carrying capacity of a compressed member is nearly always affected by buckling phenomena. The collapse modes which correspond to the ultimate limit state can be of various types:

(a) Columns of medium–low slenderness ($\lambda \leq 50$) suddenly fail without going into the large-deflection range (curve 1 of Fig. 29).

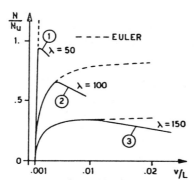

Fig. 29. Buckling curves for columns with different medium–low values of slenderness.

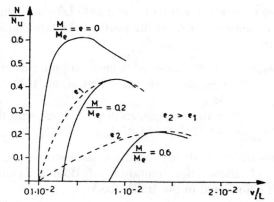

Fig. 30. Buckling curves for beam columns with medium and large values of eccentricity.

(b) Columns with high values of slenderness ($\lambda \geq 150$), when close to the collapse load, experience an elastoplastic range with small increments of the load until the limiting stress is reached in the extreme fibres. At this point, the load decreases and the phenomenon follows an unstable curve in the post-critical range, with large deformations (curve 3 of Fig. 29).

(c) Beam columns behave as in case (b), especially for large values of eccentricity (Fig. 30).

(d) Asymmetrical columns, which buckle with flexural-torsional modes, exhibit a sudden loss of load carrying capacity when torsional rotations occur.

(e) Columns in which local buckling of parts of the cross-section occurs exhibit a sudden decrease of the load dependent upon the magnitude of local buckling phenomena (Fig. 31). Curve 1

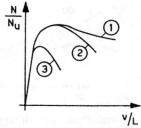

Fig. 31. Buckling curves for columns with a sudden decrease of the carrying capacity due to local buckling.

of the figure corresponds to an absence of local buckling phenomena; curve 2 represents the case in which local buckling occurs in the large-deformation range. Thus the load carrying capacity of the bar is unaffected. Curve 3, in contrast, reaches maximum load carrying capacity at values considerably less than curves 1 and 2 because of the severe interaction between global buckling of the bar and local buckling of its parts.

4.2 Ultimate behaviour of cross-sections

4.2.1 Conventional reduced elastic moment

The conventional reduced elastic moment $M_{0.2,N}$ is defined as that moment which can be transmitted through a cross-section simultaneously with a normal compression force N provided the maximum stress is not greater than $f_{0.2}$. It is assumed that the cross-section is compact and no local buckling occurs.

In the case of an extruded profile the value $M_{0.2,N}$ may be evaluated from the condition

$$\frac{N}{A} + \frac{M}{W} \leq f_{0.2} \tag{63}$$

where

N = actual normal compression force,
A = cross-sectional area,
W = section modulus.

From the above formula:

$$M_{0.2,N} = Wf_{0.2}(1 - N/(Af_{0.2})) \tag{64}$$

In the case of a beam with a transverse weld the same formula applies except that $f_{0.2}$ has to be replaced by $f_{0.2,\text{red}}$. Thus,

$$M_{0.2,N,w} = Wf_{0.2,\text{red}}(1 - N/(Af_{0.2,\text{red}})) \tag{65}$$

In the case of a beam with longitudinal welds such as shown in Fig. 5(c) the reduced cross-sectional area (eqn (31)) and the reduced section modulus (eqn (33)) have to be taken into account. From

$$\frac{N}{A_{\text{red}}} + \frac{M}{W_{\text{red}}} \leq f_{0.2} \tag{66}$$

it follows that

$$M_{0.2,N,\text{red}} = W_{\text{red}}f_{0.2}(1 - N/(A_{\text{red}}f_{0.2})) \tag{67}$$

4.2.2 Reduced plastic moment

The application of the conventional reduced elastic moment does not make use of the advantage of the strength hardening range of the material. The increase in load bearing capacity of cross-sections due to this effect is, in the case of a combined bending moment and axial force, of the same order as for pure bending. With increasing axial force the bending moment capacity of a cross-section decreases. Therefore different axial forces have different corresponding moment(M)-curvature(κ)-lines. This set of curves can be avoided by using the M–N interaction curves well known from the steel cross-sections. Such interaction curves for two types of cross-sections for aluminium girders will be presented in this section.

Along the vertical axis of the diagram the non-dimensional axial force N/N_u is shown and the horizontal axis represents the corresponding non-dimensional bending moment M/M_u. N and M are the actual internal actions corresponding to the load. The ultimate axial force N_u and the ultimate bending moment M_u will be explained later. In steel construction the linear elastic-perfectly plastic σ–ε law is valid and therefore the ultimate axial force is simply the squash load and the ultimate moment is the pure plastic moment. Transferring this to cross-sections made of aluminium the corresponding values have to be re-defined. It seems to be convenient to use the ultimate limit state values N_u and M_u respectively, which are related to the ultimate limit strain ε_{lim} as explained before. The need to fix a limit value in the most strained fibre of the cross-section and to calculate the corresponding stress σ_u via the Ramberg–Osgood law was well established in Section 3.2.3. The ultimate bending moment is

$$M_u = \int \sigma \cdot y \, dA \qquad (68)$$

and the ultimate axial force is given by the following formula:

$$N_u = \int \sigma_u \, dA \qquad (69)$$

where

σ_u = ultimate limit stress evaluated from eqn (12) with the limit strain ε_{lim}.

For two cross-sections, two aluminium alloys and different ultimate limit strains the M–N interaction curves have been evaluated by a computer simulation, and the results are presented in Figs 32(a–d), 33(a–d) and 34(a–c). As a basis for the simulation, the well known I-profile T and the hollow-profile C were taken, with in both cases strong axis bending and weak axis bending. A heat-treated and a non-heat-treated alloy were chosen. By assuming $f_{0.2} = 260$ N/mm² the first is in accordance with the 6082 alloy and the second, with a value of $f_{0.2} = 127.4$ N/mm², corresponds to the 5083 alloy. This last mentioned alloy is also representative of the heat-affected material of the heat-treated alloy 6082 within a distance of 25 mm on both sides of a weld, but with the restriction that the ultimate strain ε_t is very much lower than that for the material 5083, see also Section 3.2.3.

The non-linear increase of the σ–ε law of aluminium alloys in the strain hardening range means that limiting strain must be fixed for the calculation of the axial force and the bending moment beyond the elastic limit. If the most highly stressed fibre of the cross-section reaches this limiting strain the value N becomes $N_{u,M}$ and the value M becomes $M_{u,N}$ and they have to be calculated by integration of the appertaining stresses of all fibres (see eqn (68)). In the interaction diagrams of this section three values of limit strains ε_{lim} are defined: $5\varepsilon_{0.2}$, $10\varepsilon_{0.2}$, $0.5\varepsilon_t$ as proposed first by Mazzolani.[1] The elastic limit strain is provided by $\varepsilon_{0.2} = f_{0.2}/E$ and ε_t is the ultimate elongation at rupture usually known from the standard tensile test (see Ref. 1). The ultimate bending moment affected by the axial force N should correctly be indicated by two terms involving the axial force and the limit strain $M_{u,N,5\varepsilon_{0.2}}$, $M_{u,N,10\varepsilon_{0.2}}$, $M_{u,N,0.5\varepsilon_t}$, but the indication concerning the limit strain is omitted if the relationship between the limit strain and the corresponding ultimate moment cannot be combined. The interaction curves in this section are distinguished by the three limit strains mentioned above, consequently the corresponding relating values $N_{u,5\varepsilon_{0.2}}$, $N_{u,10\varepsilon_{0.2}}$, $N_{u,0.5\varepsilon_t}$ and $M_{u,5\varepsilon_{0.2}}$, $M_{u,10\varepsilon_{0.2}}$, $M_{u,0.5\varepsilon_t}$ have to be used. This assumption provides interaction curves intersecting the horizontal and vertical axis of the diagram at a value of 1·0.

Figure 32(a–d) show the interaction curves of the I-section T and the hollow-section C in the condition of strong and weak axis bending in the case of a heat-treated alloy. The curves representing the three limit strains mentioned before coincide over a wide range. Apart from the curve of Fig. 32(b) all the other curves have a similar shape. In the range $N/N_u \leq 0.2$ they are non-linear, but for $N/N_u > 0.2$ they are

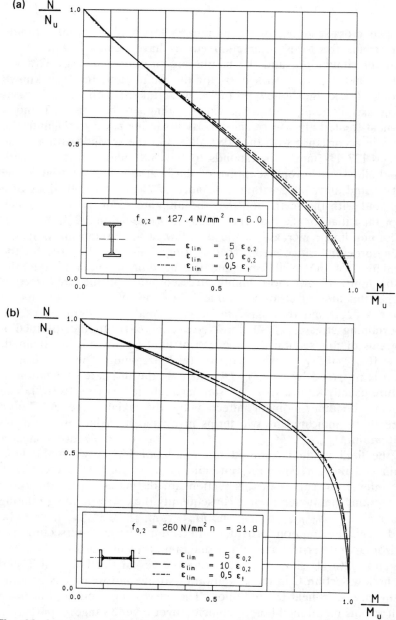

Fig. 32. Interaction diagrams for non-welded aluminium sections and different strain limits, $f_{0.2} = 260$ N/mm^2, $n = 21.8$. (a) I-section, strong axis ($f_{0.2} = 127.4$ N/mm^2, $n = 6.0$). (b) I-section, weak axis.

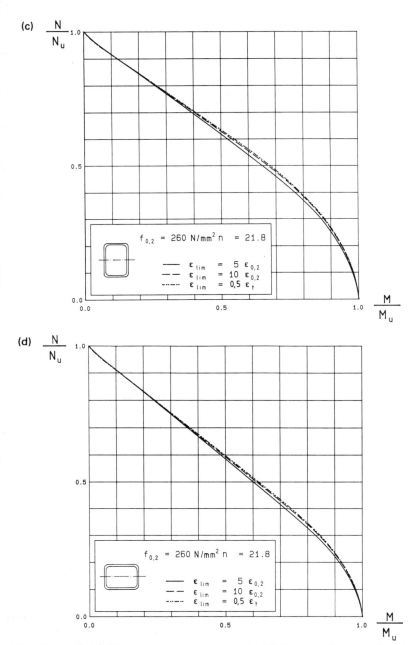

Fig. 32—*contd.* (c) Box-section, strong axis. (d) Box-section, weak axis.

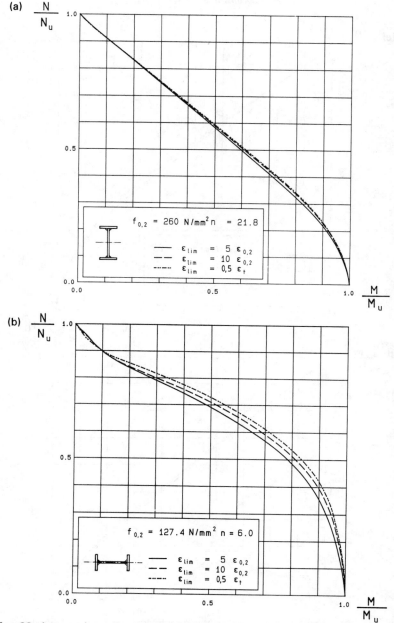

Fig. 33. Interaction diagrams for non welded aluminium sections and different strain limits $f_{0.2} = 127.4$ N/mm^2, $n = 6.0$. (a) I-section, strong axis ($f_{0.2} = 260$ N/mm^2, $n = 21.8$). (b) I-section, weak axis.

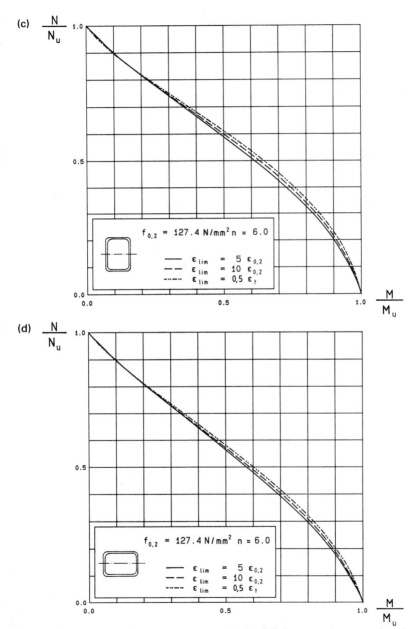

Fig. 33—*contd.* (c) Box-section, strong axis. (d) Box-section, weak axis.

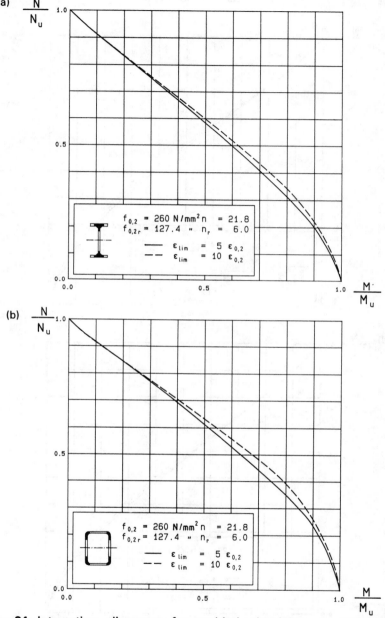

Fig. 34. Interaction diagrams for welded aluminium sections and different strain limits. Base material $f_{0\cdot2} = 260$ N/mm^2, $n = 21\cdot8$; filler material $f_{0\cdot2r} = 127\cdot4$ N/mm^2, $n_r = 6\cdot0$. (a) I-section, strong axis. (b) Box section, strong axis.

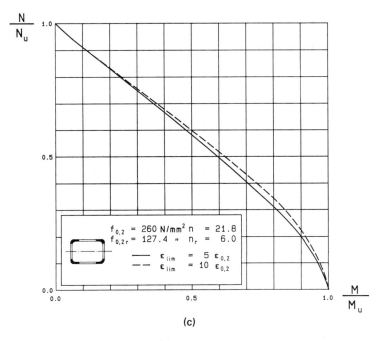

Fig. 34—*contd.* (c) Box section, weak axis.

practically linear. In Fig. 32(b) the I-profile and weak axis bending is shown. Obviously the cross-section is able to carry a certain axial load before the bending moment has to be reduced, and the behaviour is similar to the behaviour of a rectangular cross-section under the same actions. The interaction curves for the two profiles T, C and bending about the two principal axes in the case of a non-heat-treated alloy are shown in the Fig. 33(a–d). Like the curves in the Fig. 32(a–d) they practically coincide for the various limit strains. Notice that the shape of the interaction curves for the T-profile and weak axis bending (Fig. 33(b)) is similar to those in Fig. 32(b) for the heat-treated alloy, so the earlier discussion is also valid here. Altogether the interaction curves for the non-heat-treated alloy seem to be a bit less curved than those for the heat-treated alloy.

The results that are presented lead to the same conclusions as before, namely that to define an M–N interaction for profiles in aluminium:

1. By using the values $N/N_{u,\varepsilon_{\lim}}$ and $M/M_{u,\varepsilon_{\lim}}$ on the two axes of the diagrams only one interaction curve will be sufficient for all cases of limit strain conditions.
2. The influence of the strength of the alloy is negligible.
3. The interaction curves for the I-shape and weak axis bending are indeed non-linear.

After this detailed investigation it is proposed that for practical design a linear interaction formula may be used for the strength verification of a cross-section of an aluminium member under compression and bending. This interaction formula corresponds to that for steel, if the plastic values N_{pl} and M_{pl} in the steel formula are replaced by N_u and M_u. The ultimate moment $M_{u,N}$ is then given by the following equations:

$$M_{u,N} = M_u\left(1 - \frac{N}{N_u}\right)\frac{1}{0.9} \quad \text{for} \quad \frac{N}{N_u} \geq 0.1 \tag{70}$$

$$M_{u,N} = M_u \quad \text{for} \quad \frac{N}{N_u} < 0.1 \tag{71}$$

where

N = actual axial force,
N_u = ultimate axial force dependent on $\varepsilon_{\lim}$,
M_u = ultimate bending moment dependent on $\varepsilon_{\lim}$.

For simplification the values N_u and M_u may be derived from

$$N_u = A\sigma_u$$

$$M_u = (\alpha_p + \Delta_d)M_{0.2} \text{ (see Section 3.1)}$$

In the case of I-shapes and weak axis bending the $M_{u,N}$ capacity is higher and the above formulae are considerably on the safe side. An improvement may be obtained by using the following equations:

$$M_{u,N} = M_u\left[1 - \left(\frac{N}{N_u}\right)^2 \cdot \frac{1}{0.9}\right] \quad \text{for} \quad \frac{N}{N_u} \leq 0.9 \tag{72}$$

$$M_{u,N} = M_u\left(1 - \frac{N}{N_u}\right) \quad \text{for} \quad \frac{N}{N_u} > 0.9 \tag{73}$$

Members consisting of heat-treated alloys with cross welds have to be checked on the basis of the same interaction diagrams of Fig.

33(a–d) or on the basis of eqns (70) and (71). But since the ductility of the weld material and of the adjacent heat-affected material is reduced it is necessary to restrict the limit strain ε_{lim} to a maximum value of 3·5% (see Sections 3.2.3, 3.3 and 3.4.3). Consequently the curves $\varepsilon_{lim} = 0·5\varepsilon_t$ within these four figures are not applicable in the case of cross welds.

The interaction curves of cross-sections with longitudinal welds are shown in Fig. 34(a–c). Figure 34(a) is valid for the I-shape and strong axis bending, Fig. 34(b,c) are valid for the hollow section strong axis and weak axis bending. Because of the reduced limit strain in the weld the curves $\varepsilon_{lim} = 0·5\varepsilon_t$ are omitted. The ultimate value of $M_{u,red}$ and $N_{u,red}$ of cross-sections with longitudinal welds are given by eqns (68) and (69) with respect to the stress–strain law of the heat-affected material. The interaction curves are nearly equal to those of heat-treated alloys and non-welded cross-sections. Therefore it seems to be reasonable to calculate $M_{u,N,red}$ on the bases of eqns (70) and (71) as follows:

$$M_{u,N,red} = M_{u,red}\left(1 - \frac{N}{N_{u,red}}\right)\frac{1}{0·9} \quad \text{for} \quad \frac{N}{N_{u,red}} \geq 0·1 \qquad (74)$$

$$M_{u,N,red} = M_{u,red} \quad \text{for} \quad \frac{N}{N_{u,red}} < 0·1 \qquad (75)$$

For simplification the value of $N_{u,red}$ may be derived from

$$N_{u,red} = A_{red}\sigma_u$$

and

$$M_{u,red} \text{ from eqn (49)}.$$

4.3 Buckling of columns

4.3.1 Physical behaviour

Depending upon the shape of the cross-section, axially loaded bars can buckle in three different ways:

—plane buckling
—twist buckling
—lateral torsional buckling.

Plane buckling occurs when the deflection of the bar remains in a given plane. It is typical of cross-sections with two planes of symmetry in which buckling occurs in a principal plane which is:

—the minor axis bending plane, if the boundary conditions are the same in both directions,
—the plane characterized by the highest slenderness ratio if there are different boundary conditions.

If the bar does not have two planes of symmetry, plane buckling can still occur if the bar is prevented from twisting in order to avoid lateral torsional buckling.

Pure twist buckling is of major concern in those sections which have negligible warping torsional rigidity (e.g. the cruciform section), because of the convergence of the elements of the cross-section at the same point.

In this section plane buckling of extruded and welded columns is mainly considered by referring to the ECCS codification. Lateral torsional buckling is also investigated by means of computer simulation.

In the same way as for steel structures, the simulated buckling curves for simply compressed members are given in the dimensionless plane $\bar{N} - \bar{\lambda}$, with the definitions

$$\bar{N} = \frac{\sigma_c}{f_{0\cdot 2}}, \qquad \bar{\lambda} = \frac{\lambda}{\pi \sqrt{(E/f_{0\cdot 2})}} = \frac{\lambda}{\lambda_o} \qquad (76)$$

where

$f_{0\cdot 2}$ = average conventional elastic limit on the entire cross-section, or the design strength of the material (f_d),
E = Young's modulus of the material,
σ_c = failure stress, which corresponds to the load N_c that causes buckling of the bar in the considered plane,
λ = slenderness of the bar, given by $\lambda = L_c/i$,
L_c = effective length in the buckling plane, depending upon the support conditions at the ends of the bar of length L ($L_c = \beta L$),
i = radius of gyration of the cross-section in the same plane in which L_c is computed.

In order to compute the coefficient β which transforms the geometrical length into the effective one L_c, technical texts on steel structures can be referred to because the problem is mainly related to the structural systems, which are independent of the material.

4.3.2 ECCS activity

Since 1970 the ECCS committee on aluminium alloy structures has carried out extensive studies and research on the instability problems of aluminium alloy structures.[1] Several theoretical and experimental research programs have been undertaken in this field in order to investigate the mechanical properties of the materials, their imperfections and their influence on the instability of members. The numerous tests carried out have enabled a statistical evaluation of the results. These data have allowed, for the first time, the characterization of aluminium alloy members as industrial bars in accordance with the most recent trends of safety principles in metallic structures (see Section 1). These results were also extrapolated to other types of structure through a numerical analysis which makes use of computer simulation.

Among the research programs in this field, undertaken with the cooperation and support of several European countries, were the tests on extruded members carried out at Liège University.[32–34] There were also tests on welded built-up members carried out at Liège University in cooperation with the University of Naples and the Experimental Institute for Light Alloys of Novara.[35–38]

The analysis of these experimental results demonstrated the major differences between the behavior of steel and aluminium. The computational methods of the European recommendations have been based on this analysis. In particular the critical curves (a), (b) and (c) (see later) valid for bars with different cross-sections and of different materials, have been defined.

4.3.3 Theoretical analysis

The classical methods of analysis of instability problems are based upon the bifurcation of equilibrium of 'ideal bars' made of homogeneous material and having perfect geometry. In reality, because of the fabrication process which introduces 'geometrical and mechanical imperfections' into the bar (see Sections 1.3 and 1.4), the 'ideal bar' is non-existent.

The modern approach for studying instability problems is to take account of the actual imperfections present in the bar. This approach is not possible if the classical stability theory is adopted. Changing from the ideal bar to the industrial bar requires the following steps:

(1) several tests are undertaken to identify the mechanical properties and imperfections of each structural member;

(2) an analytical method is determined which allows simulation of the actual behaviour of the bar under given loads, incorporating the experimental data acquired in the previous step.

Having defined the industrial bar, the research has been primarily devoted to simulation methods because these seem to be the best methods of accounting for all of the complex phenomena involved with instability—such as the inelastic strain-hardening behaviour of aluminium alloys, the post-elastic behaviour of the members, and the effects of the welding.

Simulation methods are usually based upon numerical techniques, such as finite difference or finite element, which make use of mathematical models with fine discretizations. The most sophisticated of the numerous methods allow real structural behaviour to be closely followed.

However, it is necessary to correlate the simulation methods with the experiments that provide the input data for the numerical analysis. When the input data represent the statistical information acquired from systematic experimentation, the output of the simulation method represents an equivalent test, and therefore the analysis can be considered as semi-experimental. Thus it is necessary to make a preliminary calibration of each simulation method so that the results conform with the experimental evidence.

The simulation methods developed by Faella & Mazzolani,[39–41] by Frey[42] and by Valtinat & Müller[43,44] have been especially conceived for analysing aluminium alloy bars and were used in order to confirm the calculation methods suggested in the ECCS European recommendations.

Consider a pin-ended bar subjected to an eccentric load N in a major bending plane (Fig. 35). The analysis of the stable and unstable behaviour of the bar uses the following hypotheses, which are common to both methods:

(a) cross-sections remain plane;
(b) curvature can be approximated with the second derivative;
(c) the deformed shape of the bar remains in a given plane.

The first two hypotheses represent the classical beam theory hypothesis, and in particular the second neglects the effects of large displacements which can be considered negligible in this particular problem. The third assumption limits the deformation of the bar to a given plane.

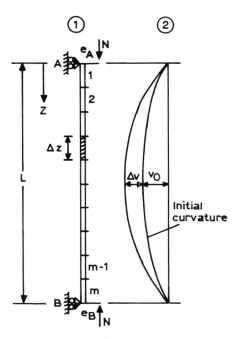

Fig. 35. Modelling of an eccentrically loaded strut for the computer simulation, major plane bending.

The Faella–Mazzolani method has the following features: from the initial deformation $v(z)$, known for the unloaded bar, each variation of load causes the deformation to change. In each step of the incremental analysis, each part of the bar is in equilibrium under the axial load

$$N = N_{\text{applied}} \tag{77}$$

and the bending moment

$$M = N(v_o + v) + Ne_A + N(e_B - e_A)\frac{z}{L} \tag{78}$$

For equilibrium:

$$N = \int_A \sigma \, dx \, dy \tag{79}$$

$$M = \int_A \sigma y \, dx \, dy \tag{80}$$

where A is the area of the cross-section of the bar referred to the principal axes x and y.

The deformation of the bar is characterized by a constant strain $\bar{\varepsilon}$ and a rotation χ about the centre of gravity. The elementary strain is therefore

$$\varepsilon = \bar{\varepsilon} - \chi y \qquad (81)$$

where, because of hypothesis (b),

$$\chi = \frac{d^2 v}{dz^2} \qquad (82)$$

The material is characterized by the Ramberg–Osgood general law (see Section 1.2.3).

The solution of the problem is given by solving eqns (77)–(82) and (12) in the six unknowns M, N, $\bar{\varepsilon}$, χ, σ and ε. These equations are solved at each load increment by a procedure which discretizes the longitudinal and transverse cross-sections of the bar and makes use of an iterative finite difference method.

Frey–Massonnet's method[42] obtains the $N-v$ diagrams by the following steps. Starting from a generic equilibrium condition, an increment similar to the previous deflection is made, varying in this way the curvature of each section of the bar. The new curvature obtained in this way is used to determine the neutral axis position by iterating until equilibrium in each section is reached. This determines variable values of N along the axis. The curvature is then changed iteratively until the value of N is constant along the bar. This method allows all the imperfections and the local elastic unloadings to be taken into account. Even though it is different from the computational point of view, it gives the same results as Faella & Mazzolani's method.

Geometrical and mechanical properties of the cross-sections used for this comparison are given in Fig. 36, and are considered with no structural imperfections because the values of $f_{0.2}$, E and n have been experimentally evaluated by means of a stub column test. An initial out-of-straightness characterized by $v_o/L = 1/1000$ has been assumed. The critical dimensionless stress $\bar{N} = \sigma_c/f_{0.2}$ is related to the slenderness ratio $\bar{\lambda} = \lambda/\lambda_o$, where $\lambda_o = \pi \sqrt{(E/f_{0.2})}$. In order to have a significant comparison, the columns have been discretized in the same way in the longitudinal direction and in the transverse direction. The comparison has been found satisfactory.[1]

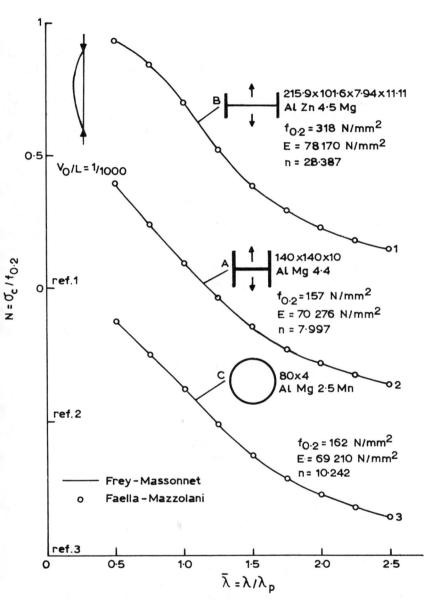

Fig. 36. Simulated buckling curves for imperfect columns with different shapes; material data.

Valtinat & Müller's method[43,44] is similar to that of Faella & Mazzolani, but it additionally takes into account two different $\sigma-\varepsilon$ laws, one for finite elements representing base material and the second for finite elements whose positions are in the weld and in the HAZ respectively.

Dangelmaier's method,[16] which has already been explained in Section 3.4.3, can also be used for plane buckling. The 'out of plane' deflections can be avoided by means of appropriate springs.

4.3.4 Experimental results

(a) Extruded profiles. Several tests on the unstable behaviour of extruded bars were undertaken by the European Convention for Constructional Steelwork (ECCS) at Liège University in 1970–72. Extruded double T profiles and box sections of different aluminium alloys from six countries were examined (see Table 12).

The following tests were carried out:[32]

(a) measurement of geometrical imperfections;
(b) chemical analysis;
(c) tension tests;
(d) determination of the distribution of Young's modulus in the cross-section;
(e) determination of the residual stress distribution;
(f) stub column test;
(g) buckling test.

Table 12. Detailed Information about Experimental Buckling Tests. Countries which Delivered, the Extruded Profiles Shapes, Dimensions and Material

Country	Shape	Dimensions (mm)		Alloy
Italy	Round tube		80 × 4	AlMg2·5Mn
Belgium	Round tube		95 × 4	AlZnMg1
France	Double T	Flanges:	140 × 140	AlMg4·5Mn
		Web:	10	
Switzerland	Double T	Flanges:	216 × 102	AlZnMg1
		Web:	8	
Switzerland	Round tube		90 × 5	AlMgSi1
Sweden	Double T	Flanges:	100 × 100	AlSiMg
		Web:	7	
Norway	Round tube		90 × 45	AlMgSi1

The load–displacement curve obtained from the stub column test describes the real behaviour of the entire cross-section under compression and takes account of all mechanical imperfections. These experimental results also confirmed that the Ramberg–Osgood law can be conveniently used to interpret the stress–strain relationship within the accuracy of experimental tolerances.

The test results have been statistically analysed to give the average stresses ($f_{c,m}$), the standard deviations (s) and the characteristic stress

$$f_{c,k} = f_{c,m} \pm ks \tag{83}$$

(see Table 13).

(b) Welded profiles. Various research programmes have been devoted to the instability of welded columns, and preliminary experimental research has been carried out on welded joints.[1] The subsequent programme on welded profiles was carried out in Italy at the ISML of Novara,[12] where residual stresses and mechanical properties were measured, and at Liège University,[8] where stub column tests and buckling tests were carried out.

Longitudinally welded members with three different cross-sections of AlSiMg alloy (6082-T6) were used (Fig. 5(c)):

Type P   double T profile built up by two extruded flanges butt welded to rolled plates (webs)

Table 13. Results from Stub Column Tests. Statistical Evaluation of Average Stress $f_{c,m}$, Standard Deviations and Characteristic Stress $f_{c,k}$

Shape	Country	Slenderness	Number of tests	Coefficient k	$f_{c,m}$ (N/mm²)	s (N/mm²)	$f_{c,k}$ (N/mm²)
Double T	France	69	9	2·306	104·2	0·868	84·6
	Switzerland	64·5	8	2·365	185·2	2·52	126·9
		48·5	8	2·365	263·2	1·51	228·1
		32·4	8	2·365	293·5	0·953	271·5
		48·3	4	3·182	255·6	0·223	194·7
Round tubes	Italy	34·4	3	4·303	137·4	0·50	116·2
	Belgium	41·15	5	2·776	270·0	0·54	255·3
	Switzerland	51·2	8	2·365	203·5	0·77	185·6
	Norway	28·9	4	3·182	199·6	2·23	194·7
		51·9	8	2·365	203·5	0·77	185·6

Type T   double T profile made of rolled plates fillet welded together in a similar way to steel cross-sections

Type C   box section made of C extruded profiles butt welded to rolled plates.

The experimental program was divided into the following parts:

(a) measurement of geometrical properties,
(b) measurement of longitudinal residual stresses,
(c) measurement of the distribution of σ–ε curves in tension and in compression in the cross-section,
(d) measurement of the specific weight,
(e) buckling tests on columns with slenderness ratios equal to 1,
(f) stub column tests.

Buckling tests have been carried out at the Laboratoire du Génie Civil in Liège, under perfectly hinged conditions. The results are given in Table 14. The average slenderness is computed using the measured lengths and the measured radii of gyration. The failure stress σ_c is computed by the collapse load N_c and the 'experimental' area A_{exp}.[1]

The experimental load displacement diagrams showed that all box section columns (C profiles, 2·70 m long) experienced progressive instability in the plane of the initial out-of-straightness.

Double T columns (P and T types, 1·1 m long) showed sudden buckling more frequently than progressive buckling owing to the small initial out-of-straightness, which led to bifurcation instability and thus to high values of the collapse load. Even though there is no relationship between the plane of buckling and the plane of initial out-of-straightness, the web eccentricity (see Fig. 3(e)) appears to compensate for the initial out-of-straightness.

4.3.5 Column buckling curve evaluation

(a) Extruded members. The method referred to here is that detailed by the ECCS Committee on a aluminium alloy structures (1977) and used in its recommendations (1978).

The buckling curves used by ECCS have been based upon the available experimental and theoretical results and the following assumptions:[45,46]

(a) A geometrical imperfection is usually assumed to be represented by an initial out-of-straightness with a midspan deflection $v_o/L = 1/1000$ (see Section 1.3).

Table 14. Results of Buckling Tests Carried Out in the Laboratoire du Génie Civil at Liège

Specimens	Slenderness (average)	Maximum load N_c (kN)	Ultimate strength $\sigma_c = N_c/A_{exp}$ (N/mm^2)	Statistical values			
				$\sigma_{c.m}$ (N/mm^2)	s (N/mm^2)	k_1	$\sigma_{c.k}$ (N/mm^2)
P11		92 000	230				
P12		95 000	238				
P14		95 000	237				
P21		88 000	221				
P22	49·06	93 000	233	226·38	10·24	2·262	202·8
P24		92 000	231				
P31		97 000	242				
P33		98 000	245				
P41		84 000	210				
P42		90 000	225				
T11		76 000	191				
T12		77 000	194				
T14		72 000	181				
T22		71 000	179				
T24	48·66	85 000	214	186·20	10·11	2·262	162·6
T31		71 000	179				
T32		73 000	185				
T33		75 000	189				
T41		76 000	191				
T42		77 000	194				
C11		93 000	207				
C21		91 000	202				
C22		86 000	192				
C31		96 000	214				
C32	48·68	95 000	211	203·84	8·06	2·306	185·2
C41		91 000	201				
C43		96 000	213				
C51		95 000	211				
C53		99 000	219				

(b) The influence of the shape of the cross-section and of its variations of thickness:
(i) is smaller than 5% and can therefore be neglected if the profile is an open symmetrical section;
(ii) cannot be neglected if the profile is an open asymmetrical section (T, U, etc.) or if it is a box section with an eccentricity equal to that defined in Section 1.3.2. It has been calculated that the decrease of strength is equal to 12% in T profiles, 13% in tubes and 16% in square box sections with an 'eccentricity'

equal to 10%. The worst case is represented by the additive action of eccentricity and asymmetry.

(c) The σ–ε law plays an important role in the shape of buckling curves, in contrast to steel structures. The Ramberg–Osgood law can be conveniently used (eqn (12)) with the values $E = 70\,000\,\text{N/mm}^2$ and $10n = f_{0\cdot 2}$ ($f_{0\cdot 2}$ in N mm^2). This assumption is approximate but has the advantage of reducing the parameters of the law to just the conventional elastic limit.

(d) Since the elastic limit of commercial aluminium alloys (for structural use) can vary from 100 to 300 N/mm^2, it seems logical to use a non-dimensionalized form for buckling curves. However, even in this case buckling curves are strongly affected by the exponent n which is related to $f_{0\cdot 2}$ as a consequence of the previous assumption. Nevertheless, experience and computation showed that the behaviour of the alloys commonly used can be mainly divided into two classes. The first comprises heat-treated alloys with large values of $f_{0\cdot 2}$ (from 200 to 300 N/mm^2), and the second non-heat-treated alloys having small values of $f_{0\cdot 2}$ (about 100 N/mm^2). Hence we can distinguish two different types of behaviour from the buckling curves.

(e) Residual stresses and variations of the elastic limit along the cross-section are negligible imperfections in extruded members (see Section 1.4.1).

(f) Since experimental results were in very close agreement with the numerical analysis, simulation computations can be systematically used to elaborate the basic data and then extrapolate them in order to produce column buckling curves for all practical cases.

Starting from these statements, the main problem was how to select buckling curves. At first glance, it seemed that the principal aspects of the problem meriting consideration were:

(a) the cross-section: box or open section;
(b) the cross-section: symmetrical or asymmetrical;
(c) the material: heat-treated or non-heat-treated alloy.

If all these variables were taken into account, too many curves would have resulted. A reduction can be obtained if the following are considered:

—the necessity to cover the commonly used alloys for structural applications;

—the advantage of having as few buckling curves as possible;
—the scatter of the buckling curves in the $\bar{N}-\bar{\lambda}$ plane from the worst to the most favourable case;
—the practical deviations of the main properties such as $\sigma-\varepsilon$ law, yield stress, shape of the profile, etc.

As a result of this study, the ECCS committee decided to adopt three non-dimensional buckling curves. (a), (b) and (c) (see Fig. 37(a)), which cover all of the extruded aluminium bars with a guaranteed elastic limit greater than or equal to 100 N/mm^2. These three curves have been computed with the following data (Fig. 37(b)):

(a) double T cross-section on the minor axis bending with $n = 20$;
(b) tubular section with an eccentricity of 10% and $n = 15$;
(c) triangular box section on the minor axis bending with an eccentricity equal to 10% and $n = 10$.

The cross-sections corresponding to curves (a) and (b) have been used in the experimental research mentioned in Section 4.4.4. The cross-section of curve (c) combines both of the unfavourable effects due to asymmetry and eccentricity.

The three curves (a), (b) and (c), drawn in Fig. 37(a) on the classical non-dimensionalized $\bar{N}-\bar{\lambda}$ diagram, are given together with the analogous ECCS curves for steel. These curves are limited by $\bar{N} = 1$, though with small slenderness higher values can be reached owing to the strain-hardening capacity of aluminium alloys.

The buckling curve to be used for a given column has to be chosen on the basis of the indications given in Table 15.

These criteria have been justified on the basis of comparisons made with experimental results and simulation analysis.[1] Since actions have been taken to practically improve the eccentricity of tubes, it was decided not to apply curve (c) at this stage until further investigations have been made.

For this reason the present ECCS Recommendations use only curves (a) and (b) (see Sections 4.5.1).

(b) Welded members. In order to emphasize the strength-reducing effects of mechanical imperfections due to welding, two types of comparison have been made on the basis of several simulation

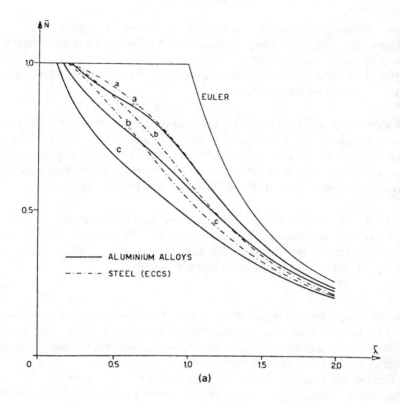

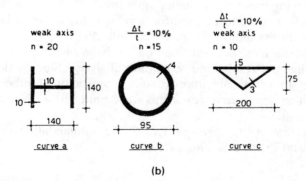

Fig. 37. Proposal for the ECCS non dimensional column buckling curves. (a) Buckling curves. (b) Basic shapes.

Table 15. Proposal for the ECCS Non Dimensional Column Buckling Curves. Indications for the Correct Choice of the Buckling Curve

Shape of cross-section	Cross-section with respect to an axis perpendicular to the bending plane	Aluminium alloy with elastic limit $f_{0.2} = 100 \text{ N/mm}^2$	
		Heat-treated	Non-heat-treated
Open	Symmetrical	a	b
Solid	Asymmetrical	b	c
Hollow	Symmetrical	b	c
	Asymmetrical	c	c

results.[36] For a given cross-section:

N = ultimate load when all the imperfections are present;
N_o = load-bearing capacity when no mechanical imperfections are present,
N/N_o gives the influence of mechanical imperfections on the load-bearing capacity.

The variation of N/N_o as a function of the slenderness ratio is given in the upper part of Fig. 38 for heat-treated alloys. The curves drawn for different cross-sections show that residual stresses have the worst effect in the $0.75 < \lambda < 1$ region. The N/N_o ratio is equivalent to the A_{red}/A ratio for $\bar{\lambda} = 0$, where A_{red} is the area of the cross-section taking account of the reduced-strength zones (see eqn (31)). This effect, which mostly affects the conventional compression capacity, tends to disappear with increasing slenderness.

This type of comparison is not of practical interest in providing a calculation method to design welded columns. It did not seem advisable to introduce further buckling curves in addition to those already defined for extruded columns, hence the same curves are used with appropriate coefficients introduced to take into account the effects of mechanical imperfections due to welding. In the case of the heat-treated profiles with double symmetry, curve (a) with the collapse loads equal to N_a had to be changed.

The ratio N/N_a represents the deviation between the effective collapse load (N) and the collapse load (N_a) given by the reference curve (a). This ratio also represents a correction factor which permits

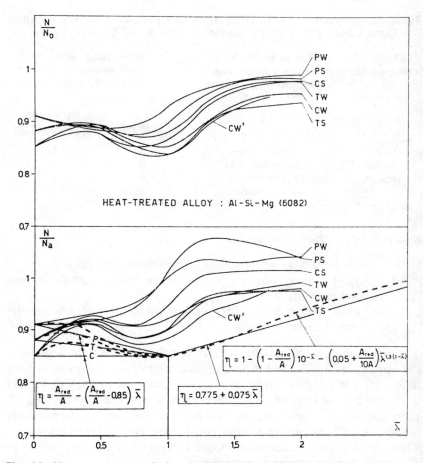

Fig. 38. Upper curves: relation between the ultimate loads of columns with longitudinal welds and without welds versus slenderness ratio for heat treated alloys. Lower curves: relation between the ultimate loads of columns with longitudinal welds and the buckling curve, a, versus slenderness ratio.

the real collapse load (N) to be obtained by multiplying it by the value from curve (a) (N_a) in each buckling case considered. The value of the correction factor has been determined by minimizing the possible N/N_a ratios (see lower part of Fig. 38).

This ratio, which is termed the reduction factor η, is equal to the ratio A/A_{red} for $\bar{\lambda} = 0$ and therefore only depends upon the shape of the cross-section. Residual stress effects are higher when $\bar{\lambda} = 1$, and the reduction factor reaches the value 0·85 which minimizes all the simulated cases. It is assumed that the buckling behaviour is independent of mechanical imperfections when $\bar{\lambda} \geq 3$; the value of η is therefore set equal to 1.

These three points ($\bar{\lambda} = 0, 1, 3$) can be connected through two lines or by an interpolating curve, which are given in Fig. 38.

In the case of non-heat-treated alloys the results of simulation have been analysed with the same criteria used for heat-treated alloys.

Since curve (b) is used for non-heat-treated extruded profiles, N/N_b ratios have to be considered in order to define the reduction factor η, which allows N values to be obtained for welded sections from curve (b). N/N_b ratios are given in Fig. 39 and in this case the T section represents the worst case, particularly when $\bar{\lambda} \cong 1$. In order to minimize the N/N_b ratios two equations can be used for the reduction factor η, in the same way that η was determined for heat-treated alloys. It can be expressed by a bilinear curve or by a continuous curve as shown in Fig. 39.

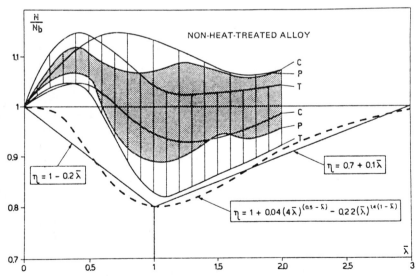

Fig. 39. Relation between the ultimate loads of columns with longitudinal welds and the buckling curve, b, versus slenderness ratio for non-heat-treated alloys.

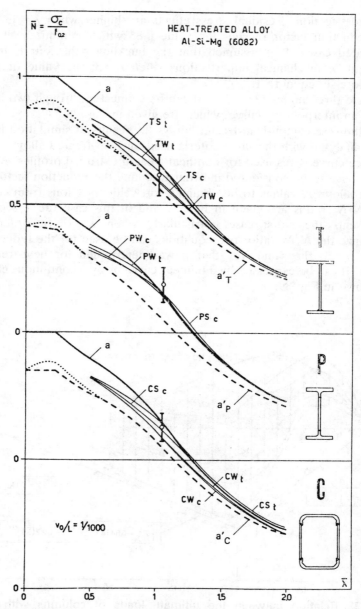

Fig. 40. Buckling curve, a, buckling loads from simulations and reduced buckling curves a' due to longitudinal welds, for different cross sections and heat treated alloys.

In the case of heat-treated alloys, the reduction factor η permits curve (a') to be obtained from curve (a) (Fig. 40). Because of the ratio A_{red}/A, curve (a') depends upon the shape of the cross-section, especially in the region $0 \leq \bar{\lambda} < 1$.

This comparison with simulation results and tests shows that the T profile represents the worst case from the experimental point of view, because of the severe geometrical imperfections of the cross-section. In fact only in this case does the characteristic value of the test results fall below curve (a'_T) by about 5%.

However, it was decided not to penalize all welded sections owing to the behaviour of T profiles because they are not the optimum choice for aluminium structures. The comparison with the simulation curves is satisfactory in all other cases.

In the case of non-heat-treated alloys, the reduction factor previously defined allows curve (b') to be obtained from curve (b). This curve (b') (see Fig. 41), in contrast to heat-treated alloys, does not depend upon the shape of the cross-section. The 18 simulation curves which represent the actual behaviour of non-heat-treated welded columns (see data of Fig. 42) are in excellent agreement with the proposed curve (b').

4.3.6 Simulation results

As explained in Section 4.3.1, the buckling mode of a compressed column depends very much on the cross-section whether it is bisymmetrical or monosymmetrical or even if it has no symmetry axis. Furthermore it is determined by the position of the load, centrally or eccentrically. Usually a strut with a bisymmetrical cross-section centrally loaded buckles with respect to the minor axis, but the introduction of initial geometrical imperfections, as described in Section 3.4.3, in the direction of the two principal axes of the cross-section ($L/1000$) and a twist around the longitudinal axis ($L/1000h$), generally provokes a lateral torsional buckling mode in a centrally loaded column with bisymmetrical cross-sections. Other imperfections such as those caused by welds may have similar influences.

The computer simulations on such members have been carried out[16] to investigate the influence of the following:

1. Different σ–ε laws for the base material (see Section 3.4.3) from the tests or based on the Ramberg–Osgood law, with E and $f_{0.2}$ from tests or given in the ERAAS.

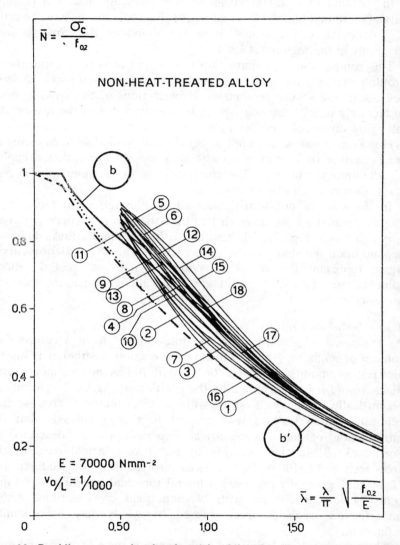

Fig. 41. Buckling curve, b, simulated buckling loads and reduced buckling curve, b', due to longitudinal welds for different cross sections and non-heat-treated alloys.

2. Cross welds and their positions along the beam column.
3. Longitudinal welds.
4. Different $\sigma-\varepsilon$ laws in the HAZ and in the weld depending on the position of the element relative to the centre of the weld (test results).
5. Depth/width/thickness ratios of bisymmetrical cross-sections.
6. Monosymmetrical cross-sections.

	RESIDUAL STRESSES	AXIS	n	σ_r (N mm^{-2})	CURVE	Nr.
T	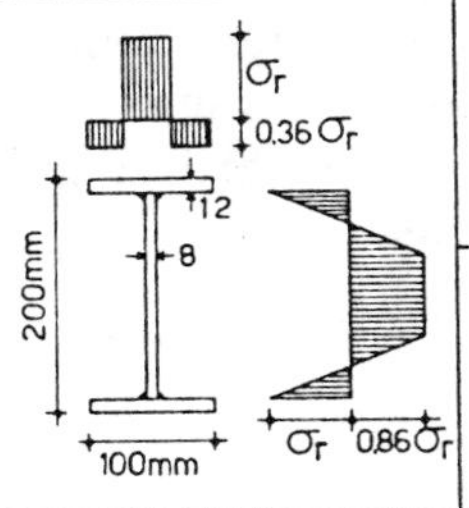	weak	12	90	TW 12	1
			16	120	TW 16	2
			20	140	TW 20	3
		strong	12	90	TS 12	4
			16	120	TS 16	5
			20	140	TS 20	6
P	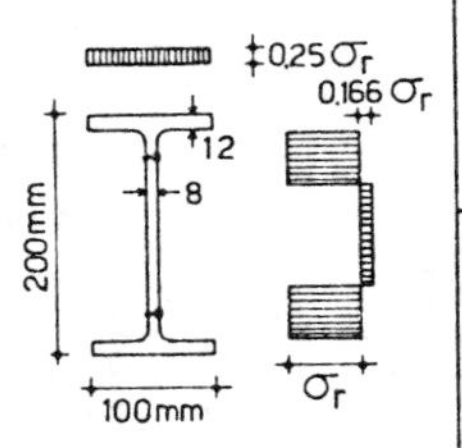	weak	12	90	PW 12	7
			16	120	PW 16	8
			20	120	PW 20	9
		strong	12	90	PS 12	10
			16	120	PS 16	11
			20	120	PS 20	12
C	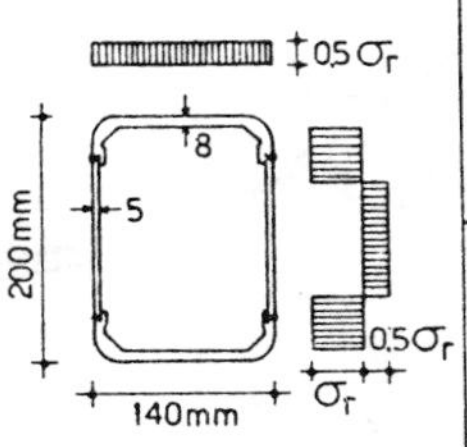	weak	12	60	CW 12	13
			16	80	CW 16	14
			20	80	CW 20	15
		strong	12	60	CS 12	16
			16	80	CS 16	17
			20	80	CS 20	18

NON-HEAT-TREATED ALLOYS

Fig. 42. Data for buckling load simulation for different cross sections with longitudinal welds.

In order to obtain information about the following problems:
1. How much do the buckling curves for extruded profiles on the basis of the Ramberg–Osgood law and E, $f_{0.2}$ from the ERAAS, differ from the buckling curves on the basis of $\sigma-\varepsilon$ laws deduced from tests (point by point)?
2. How much is the reduction in capacity with a cross weld at midspan, taking into account the different $\sigma-\varepsilon$ laws for each element corresponding to its distance from the centre of the weld?
3. What is the comparison between the buckling curve for columns with cross welds midspan (see the above No. 2) and the buckling curve for a column consisting of pure HAZ material?

Figure 43 gives the results from the computer simulations. All the following figures of this section show buckling curves with the non-dimensional buckling loads $\bar{N}$ on the vertical axis versus the slenderness ratio $\bar{\lambda}$ on the horizontal axis:

$$\bar{\lambda} = \sqrt{N_{pl}/N_{cr}} \tag{84}$$

where

$N_{pl} = A\, f_{0.2}$, squash load with $f_y = f_{0.2}$,
N_{cr} = critical lateral torsional buckling load.

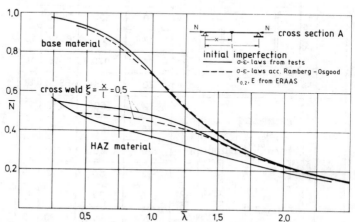

Fig. 43. Column buckling curves for non welded struts (base material), for struts with a cross weld midspan and for struts consisting purely of HAZ material. Comparison for different $\sigma-\varepsilon$-laws.

Other information about the cross-section, and its subdivision into elements, etc., may be taken from Section 3.4.3. The conclusions from Fig. 43 are:

1. There is practically no difference in the non-dimensional buckling curves using the actual σ–ε laws from tests or the Ramberg–Osgood law with E and $f_{0.2}$ from ERAAS, if it is related to the corresponding $f_{0.2}$ value of the base material.
2. The drastic reduction due to a cross weld midspan occurs for $\bar{\lambda} < 1.5$. In the range of small $\bar{\lambda}$ the relationship between the capacities for a cross welded and an extruded profile tends to β (see Section 2.2).
3. To evaluate the buckling capacity of a cross welded column by replacing all material by HAZ material is very uneconomic in the range $0.5 < \bar{\lambda} < 1.75$.

In a centrally compressed column with an initial geometrical imperfection of $L/1000$ as end levers, the position of the cross weld is of no importance since the moment gradient is nearly zero. But if such a column additionally has a uniformly distributed load with a significant moment gradient the question is:

4. How much is the buckling capacity influenced by the position of the cross weld?

The answer may be taken from Fig. 44 which also can be compared with Fig. 26:

Ans. 4. Since the normal compression force is relatively small compared to the squash load the moment gradient is significantly determined by the transverse load, so the influence of the position of the weld and the reduction in the capacity is of the same magnitude as without compression force. If the compression force is high compared with the transverse load the position of the weld has almost no influence on the buckling capacity.

The shape of the cross-section, the depth/width/thickness ratios, and the symmetry are parameters which together with welds considerably influence the buckling behaviour. In further investigations we wished to know:

5. What effect have variations in the depth/width/thickness ratios of I-struts on the buckling capacity?

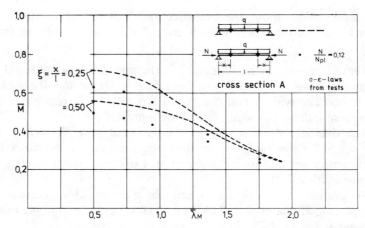

Fig. 44. Lateral buckling behaviour of a beam with cross welds at different positions. Influence of an additional longitudinal force.

6. What effect have variations in the depth/width/thickness ratio of I-struts and cross welds midspan on the buckling capacity compared with question 5?
7. Is it useful, in order to simplify the problem of evaluating the buckling capacity of cross welded struts, to replace all the material by HAZ material?

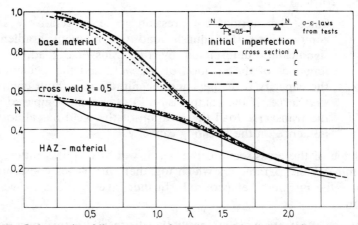

Fig. 45. Column buckling curves for non welded struts (base material), for struts with a cross weld midspan and for struts consisting purely of HAZ material; different cross sections.

The answers to these questions may be evaluated from Fig. 45.

Ans. 5. The variation in the depth/width-thickness ratios have a small influence on the buckling capacity of extruded profiles, which is about 5% in the range $0 \cdot 5 < \bar{\lambda} < 1 \cdot 0$.

Ans. 6. The influence of the depth/width/thickness ratios nearly vanishes for a column with a cross weld midspan. The drastic reduction of the buckling curves compared with those for extruded profiles has already been discussed in No. 2.

Ans. 7. It is not at all economic to replace all the material by HAZ material in order to simply evaluate the buckling loads of a column (see also the discussion in No. 3).

Evaluating the influence of the shape of the cross-section on the carrying capacity, another question is:

8. How much is the buckling load in a column with a cross weld affected if the cross section is monosymmetric, with one flange half the width of the other?

The answer may be taken from Fig. 46.

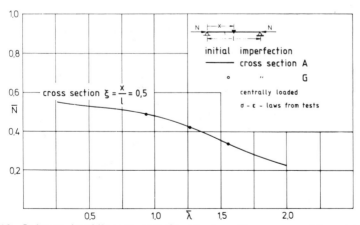

Fig. 46. Column buckling curves for struts with cross welds midspan; cross sections A and G.

Ans. 8. There is almost no influence on the buckling behaviour of a monosymmetric cross-section compared with a double symmetric one, if the initial deformation is such that the corresponding compression zone in the cross-section is on the side of the wider flange.

We now switch from columns with transverse welds to those with longitudinal welds. The investigations have been done on the same basis as before for the cross-section type A with fillet welds between the web and the flanges (see Section 3.4.3 and Fig. 21(a,b)). It is interesting to have:

9. A comparison between the buckling curves for columns of the same shape consisting of either base material, or base material and longitudinal welds, or base material and a cross weld midspan, or pure HAZ material.

The results of a computer simulation covering these problems are shown in Fig. 47.

Ans. 9. Curves ①, ② and ③ are taken from Fig. 43 (full lines) and show the buckling curves for columns consisting of:
—base material ①,
—pure HAZ material ②,
—base material and a cross weld midspan ③.
Curve ④ is computed for a column with a cross-section type

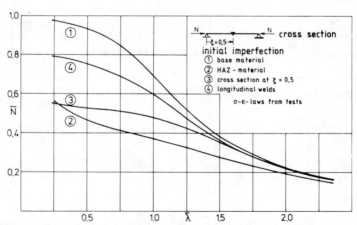

Fig. 47. Column buckling curves for non welded struts and struts with welds.

A with longitudinal fillet welds between the web and the flanges. In all cases $\bar{N}$ is related to the squash load of the full cross-section, assuming base material and taking $f_y = f_{0.2}$. The relation between the carrying capacities of curves ① and ④ tends to the value A_{red}/A for small $\bar{\lambda}$ values. It should be mentioned that if the buckling loads of curve ④ had been related to the modified $f_{0.2}$ value

$$\mod f_{0.2} = \left(\sum_i A_i f_{0.2,i}\right)/A \tag{85}$$

where each element i is combined with its corresponding $f_{0.2,i}$ value depending on its position relative to the centre of the weld, curve ④ would have nearly coincided with curve ①.

The investigations have been carried out for columns centrally loaded without and with welds. Eccentrically loaded beam columns will be treated in Section 4.4.4.

4.4 Buckling of beam columns

4.4.1 Physical behaviour

Columns can be subjected to bending caused by:

—eccentricity of the axial load,
—transverse loads.

With respect to the shape of the cross-section of the bar, buckling phenomena corresponding to bending and compression can be of two types:

—plane buckling,
—flexural-torsional buckling.

Plane buckling, which occurs in the same plane as the load eccentricity, is usually of influence in the case of open profiles with double symmetry (double T) and in the case of box sections with high torsional rigidity, where the shear centre is practically coincident with the centre of gravity.

Flexural torsional buckling is influential in the case of sections with single symmetry or without symmetry (T, C, Ω etc. or L, Z etc.) in which the position of the centre of gravity differs from that of the shear centre. The non-coincidence of the shear centre with the centre of gravity causes torsion, even in the case of axial load, which results in lateral deflection of the bar and twisting of its cross-section. However, it is also of interest to study plane buckling in these cases because there are often restraints which prevent torsional rotation of the cross-section.

4.4.2 Interaction domains

Failure conditions of a beam column are expressed by the interaction M–N curves. The first interaction formula for M–N, proposed to interpret correctly the behaviour of aluminium alloy beam columns, was that of Hill & Clark.[47,48] Under this approach it is necessary to check that

$$\frac{N}{N_c} + \frac{M}{M_u[1 - N/N_{cr}]} \leq 1 \qquad (86)$$

where

N = axial load,
M = bending moment in the case of a constant moment diagram, or equivalent bending moment in the case of different distributions,
N_{cr} = Euler critical load,
N_c = collapse load of the bar under simple compression,
M_u ultimate moment of the bar. This corresponds to the moment which causes flexural torsional buckling M_D (e.g. eqn (57)) when lateral buckling is allowed; otherwise it corresponds to the plastic moment of the cross-section M_{pl}.

Under this twofold definition of M_u, the above formula interprets both plane buckling and flexural torsional buckling.

In order to define the value of M_{pl} to be used in the case of aluminium profiles, three alternatives related to the different definitions of the conventional ultimate state of the cross-section are possible (see Section 3.2):

$$M_{pl} = \begin{cases} Wf_{0\cdot 2} & \text{(elastic limit state)} \\ \alpha_p Wf_{0\cdot 2} & \text{(plastic limit state)} \\ \alpha Wf_{0\cdot 2} & \text{(inelastic limit state)} \end{cases}$$

which give increasing values of M_{pl}, since α_p and α are the geometric and effective shape factor of the cross-section respectively (see Section 3.2.3). An analogous formula has been proposed by Massonnet for steel structures, and it is used in several specifications.

Later, in the Netherlands, TNO (1974) proposed a modified equation in order to make the design of beam columns more economic. This was based on recent experimental results on steel columns.

The following analysis was then derived and used in the ECCS steel recommendations:

plane buckling:

$$\frac{N}{N_{pl}} + \frac{\mu}{\mu - 1} + \frac{M + Ne^*}{M_{pl}} \leq 1 \qquad (87)$$

flexural torsional buckling:

$$\frac{N}{N_{pl}} + \frac{\mu_x}{\mu_x - 1} \frac{kM_x + Ne_x^*}{M_{pl,x}} \leq 1 \qquad (88a)$$

$$\frac{N}{N_{pl}} + \frac{\mu_x}{\mu_x - 1} \frac{kM_x}{M_{pl,x}} + \frac{\mu_y}{\mu_y - 1} \frac{Ne_y^*}{M_{pl,y}} \leq 1 \qquad (88b)$$

where

N = axial load,
M = bending moment or the equivalent moment,
N_{pl} = load which causes complete plasticity of the cross-section,
M_{pl} = plastic moment of the cross-section,
$\mu/(\mu - 1)$ = amplifying coefficient, with $\mu = N/N_{cr}$,
k = M_{pl}/M_D; M_D is the buckling moment (eqn (57)),
e^* = a conventional eccentricity given by

$$e^* = \left[\frac{N_{pl}}{N_c} - 1\right]\left[1 - \frac{N_c}{N_{cr}}\right]\frac{M_{pl}}{N_{pl}} \qquad (89)$$

N_c = collapse load for plane buckling of a bar under simple compression,
x, y identify the bending planes related to the corresponding variables.

The two formulations have been examined by the ECCS committee which checked these analyses against the available experimental results for aluminium beam columns (see Section 4.5.2), leading to the ECCS Recommendations (see Section 4.5.3).

4.4.3 Experimental results

A total of 167 tests have been performed, mostly in the USA and Germany, for plane buckling and for flexural torsional buckling. The formulae mentioned before and also the calculation formulae of various national codes have been checked against those tests.[1]

From the comparison between the analytical and experimental results the conclusions were:

(a) Flexural torsional buckling
   (i) The experimental results which caused flexural torsional buckling proved that the TNO formulation (eqn (88a,b)) cannot be accepted in the case of aluminium alloy beam columns since many experimental results fall into the unsafe range. The calculated values were up to 25% smaller than the experimental ones. It was generally observed that the discrepancy is higher with increasing slenderness and bending moment M.
   (ii) Clark and Massonnet's formula (eqn (86)) generally gives conservative results with only a few exceptions. The plastic range, in particular, is on the safe side. This discrepancy between calculated and experimental values is within a few percent.

(b) Plane buckling
   Equation (87) gives results that are very close to the experimental results. The deviations are smaller than those obtained by eqn (86) even though some experimental points fall in the unsafe range with deviations always less than 5%.

From the above considerations the following conclusions can be drawn:

—in the case of flexural-torsional buckling, the TNO formulation (eqn (88a,b)) is not conservative, whereas the Clark–Massonnet equation (eqn (86)) provides satisfactory results.
—in the case of plane buckling, both approaches are conservative. The TNO formulation, however, is closer to the experimental results.

For this reason the Clark–Massonnet formula has been chosen. It is safe, economic[1] and easy to handle, and therefore it has been introduced into the European Recommendations for Aluminium Alloy Structures (ERAAS, see Section 4.5.1).

4.4.4 Simulation results

In order to obtain information about the overall buckling behaviour of beam columns an extensive investigation has begun in the field of computer simulations. The projects mentioned in Sections 3.4.3 and 4.3.6 have been extended to beam columns without and with welds, and include a study of lateral torsional buckling.

For the computer simulations the same basic assumptions described in Section 4.3.6 have been made. These refer to the type of cross-section, the initial imperfections, and to welding as well as the different types of stress–strain laws (with several values for E and $f_{0.2}$).

All the diagrams of this section show buckling curves of beam columns. The bending moment is always introduced by an equal end eccentricity of the load at both ends of the bar, which produces a bending moment around the strong axis of the profile. The member always suffers lateral torsional buckling, although the lateral deflection does not sometimes start until high loads are imposed.

The vertical axes of the diagrams show the non-dimensional buckling loads

$$\bar{N} = N_u/N_{pl} \tag{90}$$

while the horizontal axes show the slenderness ratio

$$\bar{\lambda} = \sqrt{N_{pl}/N_{cr}} \tag{see 84}$$

where

N_u = the ultimate load,
$N_{pl} = Af_{0.2}$, the squash load with $f_y = f_{0.2}$,
N_{cr} = the critical lateral torsional buckling load.

The following questions were of interest and had to be answered:

1. What is the influence of a cross weld midspan for a beam column, having an eccentric load by an end lever of $e = 2k +$ initial imperfection (k = core width)? What is the influence of different $\sigma-\varepsilon$ laws? Is it useful to replace all material by HAZ material?
2. What is the influence of different eccentricities?
3. What is the influence of the combination: position of the cross weld along the length of the bar and varying geometrical imperfections?

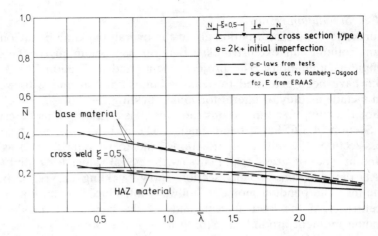

Fig. 48. Buckling curves for beam columns with cross welds midspan.

To answer these questions Figs 48–50 and Table 16 may be discussed as follows:

Ans. 1. Figure 48 shows buckling curves for beam columns with the cross-section type A and cross welds at midspan. The eccentricity of the normal force is $e = 2k +$ initial geometrical imperfection, as described in Section 4.3.6 ($k =$ core

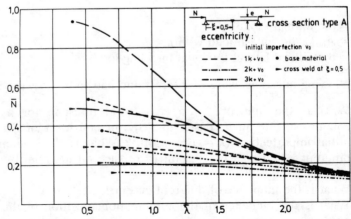

Fig. 49. Buckling curves for beam columns with cross welds midspan and different load eccentricities.

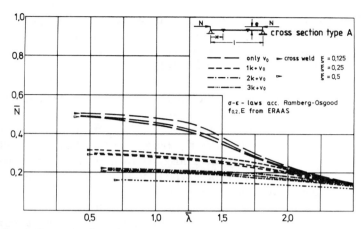

Fig. 50. Influence of end eccentricities of the loads and the position of the cross welds on the ultimate load.

width). It can be seen that there is virtually no difference between the results (full lines) on the basis of a $\sigma-\varepsilon$ law from tests (see Section 1.4.2) and those based on the Ramberg–Osgood law with E, $f_{0\cdot 2}$ taken from ERAAS (dashed lines).

The reduction in capacity between the curves for base material (case 1) and those for cross welded bars (case 2) is as expected. Since the buckling curves of members with

Table 16. Influence of Different End Eccentricities of the Loads in Comparison to the Nominal Initial Geometrical Imperfection Versus Several Slenderness Ratios

Eccentricity	Carrying capacity of the beam columns with a cross weld midspan compared with that of a beam column without welds at a slenderness ratio $\bar{\lambda} =$						
	0·5	0·75	1·0	1·25	1·4	1·75	2·0
Initial geometrical imperfection v_o	0·57	0·61	0·69	0·83	0·91	0·95	0·98
$2k + v_o$	0·59	0·65	0·69	0·73	0·78	0·83	0·85

greater eccentricities tend to be more linear (compared with typical buckling curves of centrally loaded struts) the relation between the two cases is nearly linear from about 2:1 at $\bar{\lambda}=0$ to 1:1 at $\bar{\lambda}=2\cdot25$.

The buckling curve for a bar consisting of pure HAZ material lies under the curves for welded bars, and in the range $0\cdot75<\bar{\lambda}<2\cdot25$, the reduction is considerable.

Ans. 2. Figure 49 shows buckling curves for a beam column consisting of base material (case 1) and others having cross welds (case 2) in the midspan position. The parameter is the eccentricity of the normal force, varying from the initial geometrical imperfection v_o over $k+v_o$, $2k+v_o$ to the maximum value $3k+v_o$ ($k=$ core width). It can be seen that with increasing eccentricities the buckling curves change their character to an almost straight line before they run into the Euler hyperbola. At $\bar{\lambda}=0$ the relation of the carrying capacities of bars for cases 1 and 2 is 2:1, while it becomes 1:1 in the range $1\cdot75\leqslant\bar{\lambda}\leqslant2\cdot5$. Table 16 demonstrates this relation for the two eccentricities v_o and $2k+v_o$.

Ans. 3. Figure 50 shows buckling curves for beam columns with different eccentricities of the normal forces and different positions of the cross welds along the bar. It can be seen that the lines merge into each other with increasing eccentricity. This effect has been discussed before, and the reason is that the greater the end eccentricity the less is the moment gradient along the length of the bar. In the extreme case of a constant moment, the influence of the position of the cross weld vanishes completely.

4.5 Codification

4.5.1 Columns

The buckling method, which has been codified in the first edition of ERAAS, is mainly based on the use of two fundamental buckling curves only: curve (a) applies to heat-treated alloys, curve (b) to non-heat-treated alloys (Fig. 51). In the dimensionless representation they practically cover the range of the (a,b,c) curves for steel.

Both (a) and (b) curves were derived by test results and numerical simulation for extruded columns with symmetrical cross-sections, as

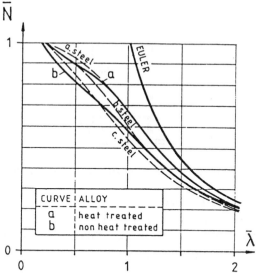

Fig. 51. European column buckling curves a, and b, for aluminium alloy structures.

has been shown in Section 4.3.5. Other cases of unsymmetrical cross-sections and welded columns are deduced from the basic curves by using appropriate reduction factors.

The load bearing capacity of a column is given by

$$N_c = K_1 K_2 \bar{N} f_d A \tag{91}$$

where

$\bar{N}$ = normalized failure load obtained by the dimensionless curves (a) or (b) as a function of the slenderness ratio,

$\bar{\lambda}$ = $\lambda/(\pi\sqrt{E/f_d})$,
f_d = design strength (assumed equal to $f_{0.2}$), $\tag{92}$
A = area of cross-section,
K_1, K_2 = reduction factors defined as follows:

Non-symmetrical cross-sections can be characterized by the unsymmetry coefficient

$$\psi = \frac{y_{max} - y_{min}}{h} \tag{93}$$

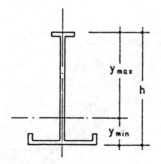

Fig. 52. Dimensions for the unsymmetry coefficient.

for y_{max}, y_{min} and h see Fig. 52. ψ can vary from 0 to 1 depending upon the rate of asymmetry ($\psi = 0$ being the symmetrical case).

The reduction coefficient K_1 which takes into account the non-symmetry of the cross-section is given by

$$K_1 = 1 - \rho\psi^2 \frac{\bar{\lambda}^2}{(1 + \bar{\lambda}^2)(1 + \bar{\lambda})^2} \tag{94}$$

where

$\rho = 2\cdot4$ in the case of heat treated alloys (curve a),
$\rho = 3\cdot2$ in the case of non-heat-treated alloys (curve b).

This empirical relationship is based upon a parametric study made through a simulation of 180 T shaped columns.[49]

In the case of members with longitudinal welds, welding effects are taken into account by means of the reduction factor K_2.[37] In the case of longitudinally welded columns, K_2 is given as follows: for heat-treated alloys,

$$K_2 = \begin{cases} A_r/A - (A_r/A - 0\cdot85)\bar{\lambda} &, \text{ for } 0 \leq \bar{\lambda} < 1 \\ 0\cdot775 + 0\cdot075\bar{\lambda} &, \text{ for } 1 \leq \bar{\lambda} \leq 3 \\ 1 &, \text{ for } 3 < \bar{\lambda} \end{cases} \tag{95}$$

where

$$A_r = A - \left[1 - \frac{f_{d,red}}{f_d}\right] \sum_i A_{rsz,i} \tag{96}$$

(reduced area, see also eqn (31)).

For non-heat-treated alloys

$$K_2 = \begin{cases} 1 - 0.2\bar{\lambda} & , \text{ for } 0 \leq \bar{\lambda} < 1 \\ 0.7 + 0.1\bar{\lambda} & , \text{ for } 1 \leq \bar{\lambda} \leq 3 \\ 1 & , \text{ for } 3 < \bar{\lambda} \end{cases} \qquad (97)$$

In the case of members with a transverse weld within the range from $0.1 L$ to $0.9 L$ the K_2 factor must be taken equal to $\beta = f_{0.2,\text{red}}/f_{0.2}$. If the welds are at the end of the bar, we can take $K_2 = 1$, but the strength must be checked.

4.5.2 Beam columns

As has been shown in Section 4.4.3, the comparison between the main theories and the existing results of tests on beam columns, where collapse occurs by excess of bending or by lateral torsional buckling, suggests that we choose the so-called Clark–Massonnet interaction formula. This is valid for bisymmetrical cross-sections, and we have to check that

$$\frac{N}{N_c} + \frac{M_{eq}}{M_u(1 - N/N_{cr})} \leq 1 \qquad (98)$$

and

$$\frac{N}{A} + \frac{\max M}{W} \leq f_d \qquad (99)$$

where

N = $\bar{N}f_d A$,
N_{eq} = equivalent bending moment,
N_{cr} = Euler critical load,
M_u = ultimate moment, which is equal to:
 M_D in the case of flexural torsional buckling
 M_{pl} in the case of plane buckling,
$\max M$ = maximum bending moment,
f_d = design strength.

In the case of biaxial bending the following check must be made:

$$\frac{N}{N_c} + \frac{M_{eq,x}}{M_{u,x}(1 - N/N_{cr,x})} + \frac{M_{eq,y}}{M_{u,y}(1 - N/N_{cr,y})} \leq 1 \qquad (100)$$

and

$$\frac{N}{A}+\frac{M_x}{W_x}+\frac{M_y}{W_y} \leq f_d \tag{101}$$

with the obvious meaning of the symbols.

In the case of longitudinal welds the same formulae can be applied provided that M_u and N_{cr} are evaluated taking account of the reduced cross-sectional data A_{red} and W_{red} and N_c from Section 4.5.1.

In the case of a cross weld it has to be assumed that $f_d = f_{0.2,red}$ without any reduction of the cross-sectional data.

4.5.3 Comparison between ERAAS and national codes

National codes. Canada: CAN 3-S 157-N 83.

The standard contains three chapters where regulations for compressed members are given. The compressive resistance C_r of axially loaded imperfect members is given in the ultimate limit state by

$$C_r = 0.9\, A F_c \tag{102}$$

There are two ranges of the buckling curve

$$\text{for} \quad \lambda \leq \lambda_o: \quad F_c = \alpha^2(1 - k\alpha\beta\lambda) \cdot F_y \tag{103}$$

$$\text{for} \quad \lambda > \lambda_o: \quad F_c = \pi^2 E/\lambda^2 \tag{104}$$

where α, k and β are coefficients, given by appropriate formulae.

The last formula shows that from λ_o on, the critical Euler load is decisive and imperfections have no influence. The break point is at a slenderness ratio of $\bar{\lambda} = 1.23$. Eccentrically compressed bars with a definite eccentricity e of the normal force, and beam columns are treated by interaction formulae which include the amplification factor

$$\frac{1}{1 - C_r/C_e} \tag{105}$$

If there is a tendency to buckle laterally, similar formulae with other coefficients are valid.

Centrally compressed members with longitudinal and partial transverse welds are treated according to the above formulae but the compressive resistance shall not be greater than

$$C_{r,w} = 0.9\, A F'_y \tag{106}$$

where

$$F'_y = \left[1 - \frac{I_w}{I}\left(1 - \frac{F_{wy}}{F_y}\right)\right]F_y \tag{107}$$

I_w = moment of inertia of the welded zone about the neutral axis,
I = moment of inertia of the total section,
F_y = yield stress of the parent metal,
F_{wy} = yield stress of the heat-affected material.

This produces a horizontal line at the resistance $C_{r,w}$ which meets the buckling curve at a certain slenderness ratio. We think that at this discontinuity point an overestimation of the carrying capacity may result.

France: Règles AL (July 1976).

The French Code is written in the ultimate limit state. The verification of the carrying capacity of compressed members is according to Dutheil's method. The stress σ due to the normal force N and the bending moment M is not allowed to go beyond the elastic limit σ_e in the most highly stressed fibre. This method has been confirmed by experimental results. The verification of the carrying capacity is given by

$$k\sigma \leq \sigma_e \tag{108}$$

The coefficient k depends on the alloy:
for non-heat-treated alloys:

$$k = \left(0.5 + \frac{0.5\sigma_e}{\sigma_k}\right) + \sqrt{\left(0.5 + 0.5\frac{\sigma_e}{\sigma_K}\right)^2 - 0.17\frac{\sigma_e}{\sigma_K}} \tag{109}$$

for heat-treated alloys:

$$k = \left(0.5 + 0.5\frac{\sigma_e}{\sigma_K}\right) + \sqrt{\left(0.5 + 0.5\frac{\sigma_e}{\sigma_K}\right)^2 - 0.8\frac{\sigma_e}{\sigma_K}} \tag{110}$$

This formula results from the fact that a straight bar has an equivalent geometrical imperfection of

$$f_1 = \beta\sigma \frac{I}{vN_K}\frac{\sigma_K}{\sigma_K - \sigma} \tag{111}$$

The factor β is 0·83 for non-heat-treated alloys and 0·2 for heat-treated alloys. I is the moment of inertia, v is the outer fibre distance, N_K is the Euler load, σ_K is the Euler stress.

The national standard of France gives detailed information on how to calculate the carrying capacity of a compressed member with longitudinal or transverse welds. Independent of the slenderness of the member the reduction of the strength due to the heat effect can be taken as zero if the weld is at the ends of the member. Furthermore it can be taken as zero if the slenderness is greater than 120. For all other cases and longitudinal welds the verification is

$$F\frac{k}{e\left[1 - 2\Sigma\, l_1(1-\beta)\left(1 - \frac{\lambda}{120}\right)\right]} \leq \sigma_e \qquad (112)$$

In this formula a reduced area due to the heat effect is taken into account. $2l_1$ is the extent of the heat-affected zone, β is the relation between the 0·2-limit of the heat-affected material and the 0·2-limit of the base material.

For other cases and for transverse welds there are two formulae for the verification:

(a) position of the weld between $0.05\,L$ and $0.1\,L$ or $0.9\,L$ and $0.95\,L$:

$$F\frac{k}{e\left[1 - \frac{1}{2}(\Sigma\, l_1 + \Sigma\, l_2)(1-\beta)\left(1 - \frac{\lambda}{120}\right)\right]} \leq \sigma_e \qquad (113)$$

(b) position of the weld between $0.1\,L$ and $0.9\,L$:

$$F\frac{k}{e\left[1 - (\Sigma\, l_1 + \Sigma\, l_2)(1-\beta)\left(1 - \frac{\lambda}{120}\right)\right]} \leq \sigma_e \qquad (114)$$

The value l_2 is the length of the transverse weld.

Germany: DIN 4113.

There are two methods of calculation for the carrying capacity of compressed members, the first one is based on interaction formulae and is applied in the ultimate limit state, the second is the traditional ω-Verfahren, which applies to the working load state. This comparison refers to the first method. The $\sigma-\varepsilon$ law is introduced as a trilinear curve which is a lower bound of a scatter band of test results (see Fig. 53).

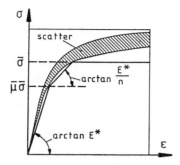

Fig. 53. Trilinear σ–ε-law in the German Code DIN 4113.

This relationship is defined by given values such as E^*, $\bar{\sigma}$, $\bar{\mu}$ and n. The interaction formulae are

$$\frac{N_v}{\bar{\mu}\bar{N}} + \frac{M_v}{\left(1 - \frac{N_v}{N^*}\right)\bar{\mu}M^*} \leq 1 \tag{115}$$

$$\psi \frac{N_v}{\bar{N}} + \frac{M_v}{\left(1 - \frac{N_v}{N^*}\right)M^*} \leq 1 \tag{116}$$

where N_v and M_v are the enhanced axial load and bending moment, the latter containing also the influence Nu, caused by a given initial imperfection u which depends on the material, the type of the profile and the slenderness.

$\bar{N} = \bar{\sigma}A$ is the elastic limit load,
$M^* = k\bar{\sigma}I/e_d$ is a factored moment of the cross-section, the factor k depends on the type of profile and on the position of the applied load,

$$\psi = 1 + \frac{n-1}{2}(1-\bar{\mu})\frac{\bar{N}}{N^* - N_v} \tag{117}$$

Both formulae are on the safe side for the whole range of slenderness, but formula (115) is more economic if

$$\frac{n+1}{2}\frac{N_v}{N^*} \geq 1 \tag{118}$$

The interaction formulae are also valid for the lateral torsional buckling of compressed members, if slight modifications are made to some values.

For compressed members with longitudinal welds the same interaction formulae can be used with the following modifications:

$$\bar{N} = \bar{\sigma} \cdot A_\kappa, \quad \text{where} \quad A_\kappa = A - (1 - f_{0\cdot 2\text{HAZ}}/f_{0\cdot 2})A_{\text{HAZ}} \quad (119)$$

$$N^* = \frac{\pi^2 E^* I_\kappa}{l^2}, \quad \text{where} \quad I_\kappa = I - (1 - f_{0\cdot 2\text{HAZ}}/f_{0\cdot 2})I_{\text{HAZ}} \quad (120)$$

$$M^* = k\bar{\sigma}_\kappa I/e_\text{d}, \quad \text{where} \quad e_\text{d} \text{ is the compressed outer fibre distance.} \quad (121)$$

Great Britain: BS 8118 (First draft, 1985).
The factored load carrying capacity of a column is given by

$$P_\text{C} = P_{\text{SC}} C_\text{C}/\gamma_\text{m} \quad (122)$$

The material factor γ_m is equal to 1·2 for riveted and bolted structures and 1·25 for welded structures. The basic axial capacity P_{SC} is given by

$$P_{\text{SC}} = \begin{cases} Af_{0\cdot 02} \text{ for extruded shapes} \\ A_\text{r} f_{0\cdot 2} \text{ for longitudinally welded shapes} \end{cases} \quad (123)$$

where A_r is the reduced area.
The reduction factor for local buckling C_L is given by

$$C_\text{L} = \frac{1}{A}\left[\sum C_{\text{EL}} + A_{\text{EL}} + \sum A_\text{x}\right] \quad (124)$$

where C_{EL} local buckling reduction factor for each element, governed by the parameter

$$\frac{hb}{t}\sqrt{\frac{A_\text{r}/Af_{0\cdot 2}}{250}} \; (f_{0\cdot 2} \text{ in N/mm}^2) \quad (125)$$

in which h values are given according to the type of reinforcement:

A_{EL} = cross-sectional area of each element,
A_x = cross-sectional area of parts of section
    not included in element width.

The reduction factor for flexural buckling C_C is governed by the slenderness parameter

$$\frac{L}{\rho}\sqrt{\frac{f_{0\cdot 2}c_k}{250}} \quad (f_{0\cdot 2} \text{ in N/mm}^2) \tag{126}$$

L being the effective length and ρ the gyration radius.

Five buckling curves for C_C values are given. The appropriate one is selected in accordance with the rules given in Table 17. It should be noted that the column buckling rules given in the first draft of BS 8118 (1985) are changed in the final draft (1992), as discussed in Chapter One.

Italy: UNI 8643.

The first Italian code for aluminium alloy structures has been issued (Dec. 1985) and it is largely inspired by the ECCS Recommendations, mainly in the buckling treatment. Both (a) and (b) curves have been accepted for heat-treated and non-heat-treated alloys, respectively. One difference from ERAAS is the evaluation of the reduction factor for welded columns, which is given by a continuous formulation:

for heat-treated alloys:

$$K_2 = 1 - \left(1 - \frac{A_r}{A}\right)10^{-\bar{\lambda}} - \left(0\cdot 05 + \frac{A_r}{A}\right)\bar{\lambda}^{1\cdot 3(1-\bar{\lambda})} \tag{127}$$

for non-heat-treated alloys:

$$K_2 = 1 + 0\cdot 04(4\bar{\lambda})^{(0\cdot 5-\bar{\lambda})} - 0\cdot 22(\bar{\lambda})^{1\cdot 4(1-\bar{\lambda})} \tag{128}$$

We also note that the design strength is different from ERAAS, and is given by

$$f_d = \min\begin{cases} f_{0\cdot 2} \\ 0\cdot 85 f_t \end{cases} \tag{129}$$

Table 17. Basic Indication for the Choice of Buckling Curve in the British Code BS 8118 (1985 draft)

Cross-section		Curve	
		$f_u/f_{0\cdot 2} \leq 1\cdot 2$	$f_u/f_{0\cdot 2} > 1\cdot 2$
Symmetric	Non-welded	1	3
	Welded	2	4
Asymmetric	Non-welded	2	4
	Welded	3	5

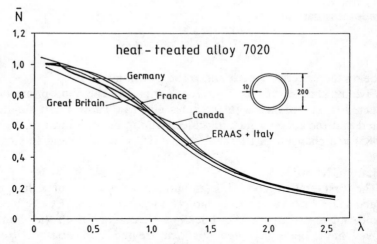

Fig. 54. Comparison of the column buckling curves of different national standards for tubes of heat treated material 7020.

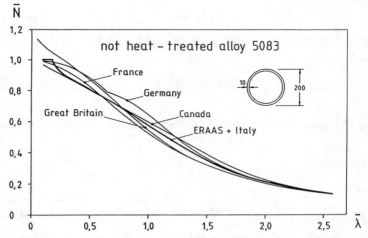

Fig. 55. Comparison of the column buckling curves of different national standards for tubes of non heat treated material 5083.

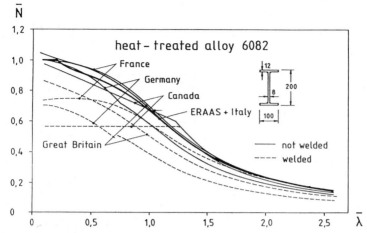

Fig. 56. Comparison of the column buckling curves of different national standards for I-sections of heat treated material 6082.

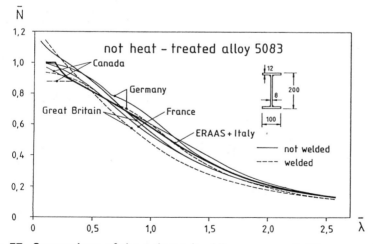

Fig. 57. Comparison of the column buckling curves of different national standards for I-sections of non heat treated material 5083.

Table 18. Design Loads of Columns with Different Cross Sections With and Without Welds According to Several National Standards

			Italy		Great Britain		Germany		France		Canada		ERAAS	
$f_{0·2}$ (N/mm²)	7 020		275		280		290		270		280[a]		280	
	6 082		245		240		260		265		260[a]		260	
	5 083		110		110		140		110		112·5		120	
	$\bar{\lambda}$		N(kN)	l(mm)	N(kN)	l(mm)	N(kN)	l(mm)	N(kN)	l(mm)	N(kN)	l(mm)	N(kN)	l(mm)
○	0·75		1 324	2 526	1 002	2 536	1 294	2 462	1 364	2 552	1 271	2 506	1 349	2 506
	1		1 102	3 369	752	3 381	1 060	3 282	1 114	3 403	1 129	3 341	1 122	3 341
7 020	1·25		841	4 210	585	4 226	789	4 104	845	4 253	988	4 177	856	4 177
○	0·75		460	3 995	394	3 998	629	3 544	438	3 998	469	3 954	502	3 828
	1		381	5 328	295	5 332	524	4 725	344	5 331	400	5 271	416	5 104
5 083	1·25		301	6 659	230	6 665	396	5 907	268	6 663	311	6 589	329	6 380
⊥	0·75		661	902	491	906	497	878	718	869	770	878	677	878
	1		518	1 203	359	1 208	407	1 170	606	1 159	685	1 170	580	1 170
6 082	1·25		399	1 504	276	1 510	312	1 463	475	1 449	599	1 463	452	1 463
⊥	0·75		252	1 348	216	1 349	376	1 196	287	1 349	307	1 334	279	1 292
	1		201	1 798	158	1 799	309	1 595	225	1 799	262	1 779	218	1 722
5 083	1·25		161	2 247	122	2 249	234	1 993	175	2 249	203	2 224	178	2 153

[a] Assumed $f_{0·2}$ values

This comparison between the national codes and the ERAAS is referred to typical profiles:
(a) round tube ($D = 200$ mm, $t = 10$ mm),
(b) I-section $200 \times 100 \times 8 \times 12$ without and with longitudinal welds.

The tube is made of a heat-treated material (7020) and a non-heat-treated material (5083), and the I-section of a heat-treated material (6082) and a non-heat-treated material (5083). To make the results directly comparable it is necessary to transform the loads into the ultimate state if codes are written in the allowable stress state. The basic data are taken from the different codes, even if they differ a little for the same material. The initial imperfections are included in each codified method, no additional bending moments are active.

The buckling curves are presented in a non-dimensional form (Figs 54–57). They show the ultimate loads versus the slenderness ratio for plane buckling (around the strong axis for the I-profile).

The dimensional values of the axial load capacity have been calculated for three slenderness ratios 0·75, 1·0 and 1·25. They are given in the table together with the corresponding length and the $f_{0·2}$ values (see Table 18).

REFERENCES

1. Mazzolani, F. M., *Aluminium Alloy Structures*. Pitman, London, 1985.
2. Baehre, R., Trydktastravorav elastoplastikt materialnagrafragestalningar. Tekn. Dr Arne Johnson Ingenjorsbyra. Report n. 16, 1966.
3. Mazzolani, F. M., La caratterizzazione della legge $\sigma-\varepsilon$ e l'instabilità delle colonne di alluminio. *Costr. Metall.*, No. 3 (1972).
4. Ramberg, W. & Osgood, W. R., Description of stress–strain curves by three parameters. NACA Tech. No. 902, 1943.
5. Sutter, K., Die theoretischen Knickdiagramme bei Aluminium-legierungen. *Technische Rundschau, Bern*, no. 20–24, 1959.
6. Mazzolani, F. M., Proposal to classify the aluminium alloy on the basis of the mechanical behaviour. ECCS, Committee 16, doc. 16-74-2, 1974.
7. Steinhardt, O., Aluminium im Konstruktiven Ingenieurbau (Aluminium Constructions in Civil Engineering). *Aluminium*, **47** (1971) 131–9, 254–61.
8. Bernard, A., Frey, F., Janss, J. & Massonnet, Ch., Recherches sur le comportement au flambement de barres en aluminium. Rapport CIDA, Liege-Paris, Mai 1971, *IABSE Mem.*, **33-I** (1973), Zürich.
9. Frey, F., Alu-alloy welded columns buckling research program: test results. ECCS-Committee 16, doc. 16-77-3, 1977.

10. Faella, C., Influenza delle imperfezioni geometriche sul comportamento instabile delle aste compresse in alluminio. *La Ricerca* (May–August 1976).
11. Mazzolani, F. M., Residual stress test on alu-alloy Austrian profiles. ECCS-Committee 16, doc. 16-75-1, 1975.
12. Mazzolani, F. M., Les imperfections structurales dans les assemblages suodées en aluminium. *Revue de l'Aluminium*, no. 431 (1974).
13. Gatto, F., Mazzolani, F. M. & Morri, D., Experimental analysis of residual stresses and of mechanical characteristics in welded profiles of Al–Si–Mg (type 6082). ECCS-Committee 16, doc. 16-77-5, 1977. *Italian Machinery and Equipment*, **11** (50) (March 1979).
14. Hill, H. N., Clark, J. W. & Brungraber, R. J., Design of welded aluminium structures. *Trans. ASCE*, **127**(II) (1962).
15. Mazzolani, F. M., Il comportamento inelastico dei profili in alluminio saldati. *Costruzioni Metalliche*, No. 5 (1971).
16. Dangelmaier, P., Traglastberechnung geschweißter räumlich belasteter Stäbe aus Aluminium. Dr.-Ing. thesis University Karslruhe, 1985.
17. De Martino, A., Faella, C. & Mazzolani, F. M., The use of plastic adaptation coefficients in the limit state design of steel shapes. *Costruzioni Metalliche*, no. 3 (1981).
18. Cappelli, M., de Martino, A. & Mazzolani, F. M., Ultimate bending moment evaluation for aluminium alloy members: a comparison among different definitions. Int. Conf. on Steel and Aluminium Structures, Cardiff, July 1987, Elsevier Applied Science, London.
19. Valtinat, G. & Dangelmaier, P., Plastic design of nonwelded and welded aluminium girders. Aluminium Weldments III. *Proc. 3rd Int. Conf. on Aluminium Weldments*, München, 1985, Aluminium–Verlag GmbH, Düsseldorf.
20. Mazzolani, F. M., Plastic design of aluminium alloy structures. Estratto dal vol. Verba volant, stripta manent Liege, 1984, pp. 295–313.
21. Valtinat, G. & Dangelmaier, P., Application of the plastic theory to nonwelded and welded aluminium members. doc. ECCS-T 2, 1984.
22. Mazzolani, F. M., Cappelli, M. & Spasiano, G., Plastic analysis of aluminium alloy members in bending: Comparison between theoretical and experimental results. Doc. ECCS-T2, October 1984.
23. De Martino, A., Faella, C. & Mazzolani, F. M., Inelastic behaviour of aluminium double-T welded beams: A parametric analysis. Doc. ECCS-T2, October 1984.
24. Lindner, J., ECCS Committee 8, Stability (in cooperation with IABSE, SSRC and CRCJ). Manual on the stability of steel structures, Introductory Report to the 2nd International Colloquium on Stability, Liège, 1977.
25. Dumont, C. & Hill, H. N., The lateral instability of deep rectangular beams. NACA Tech. Note 601, 1937.
26. Dumont, C. & Hill, H. N., The lateral stability of equal-flanged alu-alloy I-beams in pure bending. NACA Tech. note 770, 1940.
27. Hill, H. N., The lateral instability of unsymmetrical I-beams. *J. Aero. Sci.*, **9** (1942) 175.
28. Clark, J. W. & Jombock, J. R., Lateral buckling of I-beams subjected to unequal end moments. *J. Engng. Mech. Div., ASCE*, **EM3** (1957).

29. Clark, J. W. & Rolf, R. L., Buckling of aluminium columns, plates and beams. *J. Struct. Div., ASCE,* **St3** (1966) 17.
30. Klöppel, K. & Bärsch, W., Versuche zum Kapitel Stabilitätsfälle der Neufassung von DIN 4113. *Aluminium,* No. 10 (1973) 690–9.
31. Frey, F., Buckling, lateral buckling and eccentric buckling of alu-alloy columns, beams and beam-columns. ECCS-Committee 16, Doc. 16-77-1, March 1977.
32. Bernard, A., Etude sur le flambement des barres industrielles en aluminium (Study on buckling of aluminium industrial bars). ECCS Committee 16, doc. 1.1-73-3, February 1973.
33. Flaiaux, M. & Frey, F., Etude d'une serie de problèmes posés par le flambement centré plan des colonnes industrielles en alliages d'aluminium (Study of a series of problems for plane buckling of aluminium alloy industrial columns). Travail de Fin d'Etudes, University of Liège, (1974).
34. Frey, F. & Rondal, J., Aluminium alloy buckling curves a,b,c: Table and equations. ECCS Committee 16, doc. 16-78-1, March 1978.
35. Mazzolani, F. M., The influence of mechanical imperfections on the structural behaviour of welded aluminium alloy members. 2nd Int. Conf. on Aluminium Weldments, Munich, May 1982, Aluminium-Verlag, Düsseldorf.
36. Faella, C. & Mazzolani, F. M., Buckling behaviour of aluminium alloy welded columns. ECCS Committee 16, doc. 16-78-2, June 1978.
37. Faella, C. & Mazzolani, F. M., European buckling curves for aluminium alloy welded members. *Alluminio,* No. 11 (1980).
38. Mazzolani, F. M., Welded construction in aluminium. European Recommendations: Welded members. *Proc. I.I.W. Colloquium on Aluminium,* Porto, September 1981.
39. Faella, C. & Mazzolani, F. M., Simulazione del comportamento di aste industriali inelastiche sotto carico assiale (Simulation of the behaviour of in-elastic industrial bars under axial load). *Construzioni Metalliche,* No. 4 (1974).
40. Faella, C. & Mazzolani, F. M., Sul comportamento post-critico di aste sotto carichi generici (On the post-buckling behaviour of bars under generical loads). 2nd National Congress AIMETA, Naples, October 1974.
41. Mazzolani, F. M., Inelastic buckling of metal bars. Stavebnicky Casopis, September 1975.
42. Frey, F., Calcul de flambement des barres industrielles (Buckling computation of industrial bars). *Bulletin Technique de la Suisse Romande,* No. 11 (May 1971).
43. Valtinat, G. & Müller, R., Alu-Alloy welded column buckling research program: numerical computations. ECCS Committee 16, Doc. 16-76-3, 1976.
44. Valtinat, G. & Müller, R., Ultimate load of beam-columns in aluminium alloy with longitudinal and transversal welds. *Proc. 2nd Int. Colloquium on Stability of Steel Structures,* Liège, April 1977, European Convention for Constructional Steelwork, Brussels.
45. Mazzolani, F. M. & Frey, F., Buckling behaviour of aluminium alloy extruded members. *Proc. 2nd Int. Colloquium on Stability of Steel*

Structures, Liège, April 1977, European Convention for Constructional Steelwork, Brussels.
46. Faella, C. & Mazzolani, F. M., European buckling curves for aluminium alloy extruded members. *Alluminio,* No. 10 (1980).
47. Hill, H. N. & Clark, J. W., Lateral buckling of eccentrically loaded I-section columns. *ASCE Trans.,* **116** (1951) 1179.
48. Hill, H. N. & Clark, J. W., Lateral buckling of eccentrically loaded I-section and H-section columns. *ASCE Proc. First US Nat. Congr. Appl. Mech.,* 1951, p. 407.
49. Frey, F., Flambement plan des Colonnes industrielles a section droite asymetrique. Verba volant, scripta manent, Liège, 1984, pp. 295–313.

3

Lateral-Torsional Buckling of Beams

D. A. NETHERCOT

University of Nottingham, Nottingham, UK

ABSTRACT

Data on the lateral-torsional buckling of aluminium beams of open cross-section are reviewed within the context of their usefulness as the research base for codified design approaches. Particular attention is given to test data and ultimate strength theoretical solutions. The simplified approach of the British BS 8118 is explained and detailed comparisons with the research base presented. In addition to the positioning of the basic design curve, attention is given to the treatment of non-uniform moment loading, monosymmetry of the cross-section and the interaction with loss of effectiveness due to local buckling effects in slender cross-sections.

NOTATION

b	Flat width of plate element
B	Overall width of section
c	Buckling curve parameter or lip depth
D	Overall depth of section
E	Young's modulus
h	Distance between flange centroids
I_w	Warping constant
I_x	Major second moment of area
I_y	Minor second moment of area

I_{yc}	Second moment of area of compression flange about minor axis
J	Torsion constant
k	Effective length factor
k_L	Coefficient of plate effectiveness
L	Length
m	Equivalent uniform moment factor
M	Applied moment
$\bar{M}$	Equivalent uniform moment
M_b	Lateral-torsional buckling moment
M_c	Moment capacity of cross-section
M_E	Elastic critical moment
n	Factor used to define material stress–strain curve
$p = I_{yc}/I_y$	
p_o	Material design strength
$p_1 = M_c/S_n$	Reduced material design strength, see eqn (8)
q	Buckling curve parameter
r_y	Radius of gyration about minor axis
S_n	Plastic section modulus
t	Web thickness or plate thickness
T	Flange thickness
y_o	Distance between centroid and shear centre
X, Y	See eqn (1) and Table 3
Z_e	Effective elastic section modulus for a slender section
Z_n	Elastic section modulus
$\beta = f(b/t)$	Plate buckling parameter, see eqn (6)
β_o, β_1	Fully compact, semi-compact limits
β_m	Ratio of end moments
β_x	Cross-sectional parameter, see Fig. 18
δ_x, δ_y	Amplitude of initial imperfection
$\varepsilon = \sqrt{250/p_o}$	
η	Buckling curve imperfection
θ_z	Amplitude of initial twist
$\bar{\lambda}_1$	Plateau length on buckling curve
$\bar{\lambda} = \sqrt{M_c/M_E}$	for fully compact section
$\bar{\lambda} = \sqrt{p_o S_o/M_E}$	for semi-compact or slender section
$\sigma_{0.2}$	0·2% proof stress
σ_{ult}	Material ultimate stress

1 INTRODUCTION

Thin-walled aluminium members of open cross-section bent about their major axis are prone to failure by lateral-torsional buckling[1] as illustrated in Fig. 1. Design codes therefore need to include a reasonably simple means of relating the load level at which this will occur to the physical description of the problem, i.e. beam geometry, load type, support conditions, etc. This is most conveniently expressed in terms of a beam design curve. Whilst such curves are likely to be similar to those adopted elsewhere for steel beams, differences in the details of the response of steel and aluminium members mean that the exact positioning must be based on data that relates specifically to aluminium beams.

The European Recommendations for both steel[2] and aluminium[3] structures use a re-ordered Merchant–Rankine type of beam design curve, first proposed by Trahair,[4] in which the parameter q controls the exact position of the curve. Section 3 of Chapter 2 describes this more fully. As an alternative, recent British Standards for the structural use of steel[5,6] and aluminium[7] use a modified form of the Perry–Robertson approach, of the type first proposed for struts by Dwight,[8] in which the form of the notional imperfection η is used to locate the curve. In all cases the framework within which the design

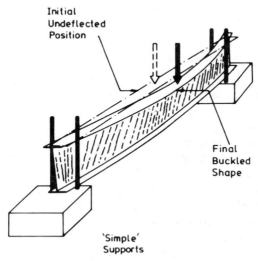

Fig. 1. Lateral buckling of an I-beam.

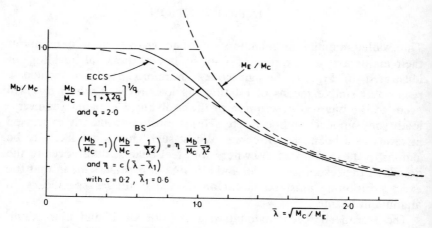

Fig. 2. Design curves for laterally unrestrained beams.

curve is specified is one of relating the buckling resistance moment M_b, suitably non-dimensionalised using a measure of the cross-section's moment capacity M_c, to some form of non-dimensional slenderness parameter $\bar{\lambda}$. Figure 2 compares examples of both types of curve with the upper bound given by the lesser of M_c or the theoretical elastic critical moment M_E.[1]

Two possibilities exist for use as the basis for the exact positioning of the design curve:

(i) test data,
(ii) accurate numerical results.

In the case of aluminium beams, Frey[9] has assembled some 90 tests from six different sources and has used them to justify a value for q of 2·0 for aluminium beams. However, most of these tests were conducted many years ago, several were on beams of narrow rectangular cross-section (for which warping effects are negligible), whilst in each case only very limited information is available on alloy properties, cross-sectional dimensions, etc. Thus differentiation between different classes of member in the way that has been accepted for struts in both steel[10] and aluminium[11] is not really possible. The same sets of test data have also been used as the basis for the beam curve in the new UK aluminium code.[7]

An approach which is wholly reliant upon the use of test data has several disadvantages, the most severe of which is that the data can only provide an uneven and incomplete coverage of the key parameters. Providing a numerical simulation has been properly verified—usually by a series of one for one checks against the best available test data—it may be used in a systematic way to produce results which cover reasonable variations in all the key problem parameters. Although such a capability has been available for some time for aluminium struts,[12] it has only recently been developed for the more complex flexural-torsional behaviour of aluminium beams.[13] In this context it is, of course, necessary that the analysis be based on a true ultimate strength approach and not simply be of the bifurcation type. In this way the full effects of imperfections on the growth of out-of-plane deformations with increasing applied load may be properly followed.

Figures 3–6 illustrate the effects of selected key parameters on the lateral-torsional buckling strength of aluminium beams. In all cases the results are as might be expected. Reductions in strength caused by changing the section shape via an increase in the D/T ratio and increasing the level of initial out-of-straightness δ_y and/or θ_z agree qualitatively with the behaviour of steel beams. Changing the material stress–strain curve by either increasing $\sigma_{0.2}$ or n produces effects that are in line with the behaviour of aluminium struts, including the

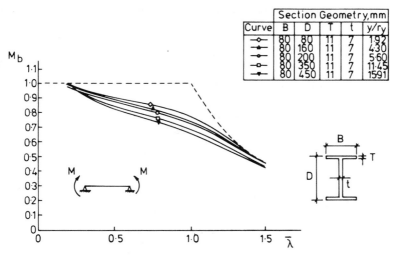

Fig. 3. Effect of section geometry.

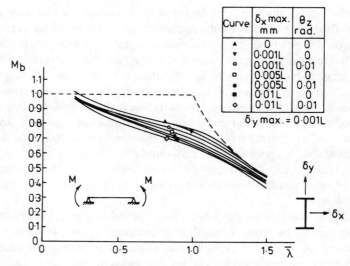

Fig. 4. Effect of initial out-of-straightness.

cross-over of results observed at low values of n. Figure 7 shows that a change from uniform moment loading to a moment gradient causes the results to plot progressively higher as already established for steel beams.[14] It is also possible to use the results of Figs 3–7 to estimate the effects of changes to combinations of parameters by noting that each

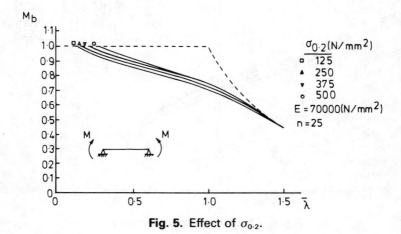

Fig. 5. Effect of $\sigma_{0 \cdot 2}$.

Lateral-torsional buckling of beams

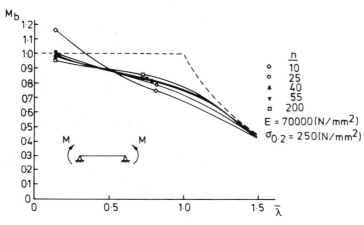

Fig. 6. Effect of n.

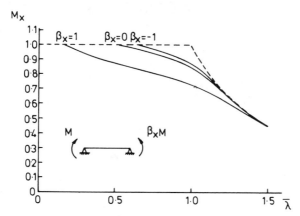

Fig. 7. Buckling of non-welded beams under unequal end moments ($\beta_x = 1, 0, -1$).

Table 1. Basic Problem Parameters for Results of Figs 3–7

Cross-section	Other parameters
$D = 200$, $T = 11$, $t = 7$, $B = 80$ (mm)	$E = 70\,000$ N/mm^2 $\sigma_{0\cdot 2} = 250$ N/mm^2 $n = 25$ $\delta_x = L/1\,000$ $\delta_y = L/1\,000$ $\theta_z = 0\cdot 01$ radians Ends simply supported Uniform moment loading

set of results contains a common curve for the basic problem of Table 1.

2 BASIS OF UK DESIGN PROCEDURE

As indicated earlier, the approach adopted for the design of laterally unrestrained beams in the new UK aluminium code[7] follows broadly that taken previously in recent steel codes. Thus the framework of Fig. 2 forms the basis, with M_b and $\bar{\lambda}$ being related as shown.

Positioning of the design curve is controlled through the η-parameter and Fig. 8 compares different choices. That finally selected

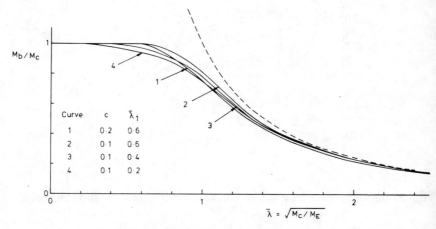

Fig. 8. Comparison of design curves.

Table 2. Test Data for Laterally Unrestrained Beams

Ref	Authors	No. tests	Cross-section	Comments
15	Kloppel/Barsch	8	I	Tip loaded cantilevers
16	Dumont/Hill	24	I	Pure bending
17	Clark/Rolf	9	I	Pure bending
18	Dumont/Hill	9	I	Pure bending, some end restraint
19	Clark/Jombock	37		
20	Hill	4	I	$\beta = 1, 0.5, 0, -0.5, -1.0$, some end restraint
			I	Pure bending, unequal flanges, some end restraint

is compared with the test data for doubly symmetrical sections (I's or flats) under uniform moment from the set of Table 2 (53 results) in Fig. 9. For values of $\bar{\lambda}$ in excess of unity the curve is clearly an approximate lower bound—although the amount of scatter in the results is quite modest in this region—whilst for more stocky beams several tests points fall below the curve. The mean value of $M(\text{test})/M(\text{predicted})$ for all the results is, however, slightly greater than 1·0. A very similar picture is obtained if the ECCS design curve of Fig. 2 is used, even though this is slightly lower for $0 \cdot 2 \geqslant \bar{\lambda} \geqslant 1 \cdot 2$.

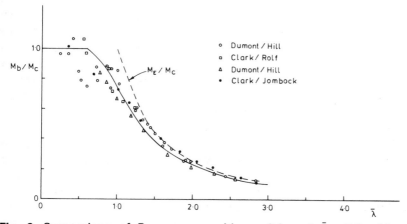

Fig. 9. Comparison of Perry curve with $c = 0 \cdot 2$ and $\bar{\lambda}_1 = 0 \cdot 6$ with all available test data for uniform moment loading.

The evaluation of $\bar{\lambda}$ is facilitated through a combination of algebraic rearrangement and approximation of some cross-sectional properties. Thus $\bar{\lambda}$ may be written as:

$$\bar{\lambda} = \sqrt{M_c/M_E}$$
$$= \frac{\lambda}{\pi}\sqrt{\sigma_{0.2}/E} \qquad (1)$$

with λ being taken as one of:

$\lambda = L/r_y$ —safe approximation for rapid calculation

$\lambda = \dfrac{X(L/R_y)}{\left[1 + Y\left(\dfrac{L/r_y}{D/T}\right)^2\right]^{1/4}}$ —suitable for certain standard shapes for which expressions for X and Y may be derived

$\lambda = \pi\sqrt{ES_n/M_E}$ —general expression, in which S_n is the plastic section modulus of the cross section

Table 3 lists suitable expressions for X and Y for plain and lipped Is and channels.

3 COMPARISON OF BS DESIGN CURVE AND THEORETICAL RESULTS

The availability of numerical results of the sort presented in Figs 3–7 permits a separate assessment of the design curves of Fig. 2 in terms of the influence of changes to single parameters. For example Fig. 10 compares the BS design curve with the results of Fig. 3 for changes in material stress–strain behaviour, i.e. different alloy groups. Noting that the results for $n = 25$ appear on each of Figs 4–7 permits an approximate comparison to be made between the relative positions of the design curve and all the numerical results.

The comparison with the basic problem of Table 1 confirms the general suitability of the design curve for values of $\bar{\lambda}$ in excess of about 0·9 but suggests that it is unconservative by about 10% for more stocky beams. Varying the beam parameters, either singly or in combination, to arrive at a higher D/T ratio, larger initial out-of-straightness, a lower value of $\sigma_{0.2}$ or a lower n value for low slenderness or a higher n value for high slenderness would suggest a

Table 3. Approximations for X and Y (see eqn (1))

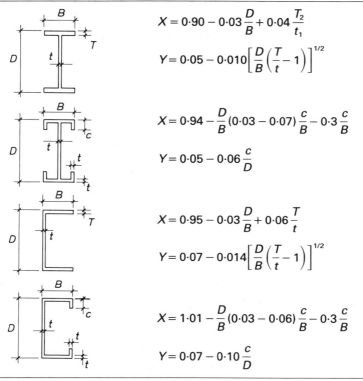

For I-section:
$$X = 0.90 - 0.03\frac{D}{B} + 0.04\frac{T_2}{t_1}$$
$$Y = 0.05 - 0.010\left[\frac{D}{B}\left(\frac{T}{t} - 1\right)\right]^{1/2}$$

For monosymmetric I-section:
$$X = 0.94 - \frac{D}{B}(0.03 - 0.07)\frac{c}{B} - 0.3\frac{c}{B}$$
$$Y = 0.05 - 0.06\frac{c}{D}$$

For channel:
$$X = 0.95 - 0.03\frac{D}{B} + 0.06\frac{T}{t}$$
$$Y = 0.07 - 0.014\left[\frac{D}{B}\left(\frac{T}{t} - 1\right)\right]^{1/2}$$

For lipped channel:
$$X = 1.01 - \frac{D}{B}(0.03 - 0.06)\frac{c}{B} - 0.3\frac{c}{B}$$
$$Y = 0.07 - 0.10\frac{c}{D}$$

Note 1. The expressions for X and Y are valid for: $1.5 \leq D/B \leq 4.5$, $1 \leq t_2/t_1 \leq 2$, $0 \leq c/B \leq 0.5$.

2. For the specific shape of lipped channel standardised in BS 1161 $X = 0.95$, $Y = 0.71$.

greater degree of unconservatism for stocky beams and the appearance of some lack of conservatism for slender beams.

A full explanation for this trend for the mean of the test results to plot higher than the mean of the numerical results over part of the range has yet to be provided. The program used to produce the latter has been widely checked against test data for lateral-torsional buckling of aluminium beam-columns,[13] as well as against several series of test data and alternative numerical results for steel members.[21,22] In a similar exercise for steel beam-columns Lindner & Gietzelt[23] found that their numerical results also fell consistently below the spread of

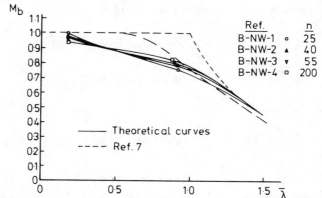

Fig. 10. Comparison of BS design curve with numerical results $\left(\dfrac{\sigma_{ult}}{\sigma_{0\cdot 2}} \leqslant 1\cdot 4\right)$.

the test data, whilst recent work on end restrained members[24] shows how small changes in the degrees of end restraint can sometimes have large effects on member strength. This point is of particular significance for the present test data, as Table 2 shows that several series were conducted in such a way that partial end restraint was present. Plotting of the test points has in all cases been done using an effective length correction based on elastic behaviour. It is well known[25,26] that for stocky members buckling in the inelastic range, yielding will reduce main member stiffness with the result that larger corrections become appropriate, i.e. points should move horizontally towards the origin. The amount of such a shift is, however, difficult to determine. To illustrate the importance of this point Fig. 11 shows the uniform moment test data of Ref. 19 plotted using both the uncorrected geometrical slenderness and values obtained by assuming an effective length of $0\cdot 55L$, this figure being the average correction suggested by Clark and Jombock. It is also noticeable from Fig. 9 that the vertical spread of the test points increases significantly for the more stocky beams, since the positioning becomes more dependent on the M_c value, which in turn depends upon an accurate knowledge of $\sigma_{0\cdot 2}$. This was not always available on an individual test basis. Furthermore for $n \geqslant 25$ the material σ–ε curve is quite flat at high strains with the result that a

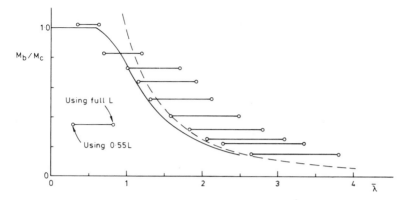

Fig. 11. Effect of effective length factor used in plotting test data, uniform moment results of Ref. 19.

moment value of $\sigma_{0.2}S$ can only be approached asymmetrically; the appearance of test points above the line corresponding to $M_b/M_c = 1\cdot0$ therefore suggests the presence of some variability in material characteristics.

Taking an overall view of both the test data and the numerical results as the basis for positioning a design curve for laterally unrestrained beams suggests that both the ECCS and BS curves are satisfactory for the more slender range, say $\bar{\lambda} > 0\cdot9$, but that both may be considered slightly unconservative for stocky beams. In this context it is worth noting that in practice when using the BS design method as a whole $\bar{\lambda}$ values are often likely to be somewhat overestimated due to the approximate determinations permitted, whilst small degrees of unquantifiable end restraint from practical connections will also cause the $\bar{\lambda}$ values used in calculations to tend to be on the high side.

4 EFFECT OF NON-UNIFORM MOMENT

It is well known that beams subject to a linearly varying moment as illustrated in Fig. 12 are inherently more stable than the equivalent beam when loaded by equal end moments ($\beta_m = 1\cdot0$). Basic elastic buckling theory[1] suggests a convenient approach to this in design as being through the concept of 'equivalent uniform moment factors', i.e.

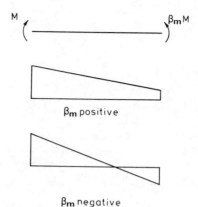

Fig. 12. Moment gradient loading.

replacing, for the purpose of the lateral buckling check, the actual moment pattern with uniform moment of a lesser value.

$$\bar{M} = mM \quad (2)$$

For elastic buckling the appropriate values of m for $1 > \beta_m \geq -1$ are virtually independent of all factors other than β_m, being only slightly affected by beam geometry, slenderness and symmetrical end restraint.[1] Various expressions linking m and β_m have been proposed; the simplest is

$$m = 0.6 + 0.4\beta_m \not< 0.4 \quad (3)$$

whilst that used in the UK steel code[5] is

$$m = 0.57 + 0.33\beta_m + 0.10\beta_m^2 \not< 0.43 \quad (4)$$

Figures 13–16 compare the test results from Ref. 19 for $\beta_m = 0.5$, 0.0, -0.5 and -1.0 with the BS design curve modified using the appropriate m factors from eqn (4). For the purpose of the comparisons the design curve has been multiplied by $1/m$ up to a limit of $M_b/M_c = 1$; in the Code the approach is to check M_b determined in the ordinary way against $\bar{M}$ and to ensure adequate cross-sectional strength at the end with the larger moment by making a separate check that $M \not> M_c$.

In all cases the comparisons of Figs 13–16 show that the use of m factors provides a simple means of predicting the test strengths with

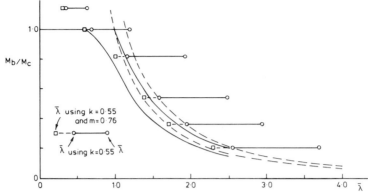

Fig. 13. Comparison of design curves with test data for $\beta_M = 0.5$.

approximately the same degree of accuracy as exists between the basic design curve and the test data for $\beta_m = 1.0$. Moreover the same trend of generally safe predictions for slender beams and some instances of unconservative predictions for stocky beams may be observed.

An alternative method of dealing with the effect of moment gradient in design consists of modifying the beam's slenderness $\bar{\lambda}$ by using enhanced values of M_E in eqn (1) that reflect the actual loading, i.e. using M_E/m if M_E is the value for $\beta_m = 1.0$. This approach is used for example by the ECCS.[2,3] It has been applied to the $\beta_m = 1.0$ results of Ref. 19 and is shown compared against the BS design curve in Figs

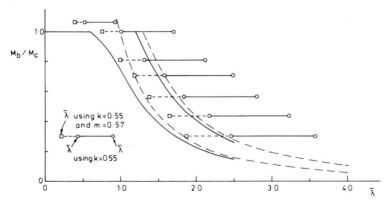

Fig. 14. Comparison of design curves with test data for $\beta_M = 0.0$.

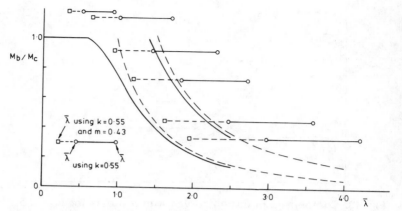

Fig. 15. Comparison of design curves with test data for $\beta_M = -0.5$.

13–16. In all cases the test results are conservatively predicted. Whilst this could be advanced as a strong argument for accepting the 'slenderness correction' approach in preference to the m-factor approach, an important element in using the latter is its suitability as a simple method of dealing with moment gradient loading on beam columns.

The suitability of the m-factor approach has also been compared with the numerical data for $\beta_m \neq 0$ from Ref. 13. Figure 17 shows, on the basis of scaling the numerically obtained $\beta_m = 1 \cdot 0$ curve by the

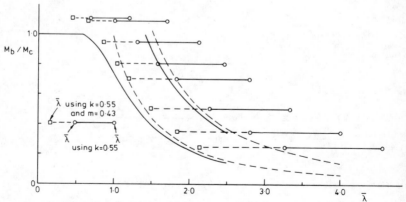

Fig. 16. Comparison of design curves with test data for $\beta_M = -1 \cdot 0$.

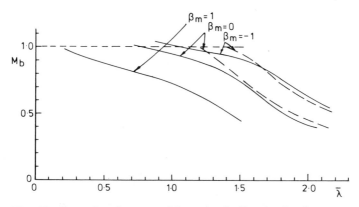

Fig. 17. Use of m factors with numerically obtained curves.

appropriate factors from eqn (4), that the results for non-uniform moment may be accurately predicted using such an approach.

5 UNEQUAL FLANGED BEAMS

The elastic buckling of beams that are symmetrical about the minor axis only[1] is complicated by the non-coincidence of the section's centroid and shear centre. Thus the basic expansion for M_E becomes

$$M_E = \left(\frac{\pi^2 EI_y}{L^2}\right)^{1/2} \left[\left(GJ + \frac{\pi^2 EI_w}{L^2} + \left\{\frac{\beta_x}{2}\left(\frac{\pi^2 EI_y}{L^2}\right)^{1/2}\right\}^2\right)^{1/2} + \frac{\beta_x}{2}\left(\frac{\pi^2 EI_y}{L^2}\right)^{1/2}\right] \quad (5)$$

in which

EI_y = minor axis flexural rigidity,
GJ = torsional rigidity,
EI_w = warping rigidity,
$\beta_x = \left[\dfrac{\int x^2 y \, dA + \int y^3 \, dA}{I_x} - 2y_o\right]$
I_x = second moment of area about the major axis,
y_o = distance between centroid and shear centre.

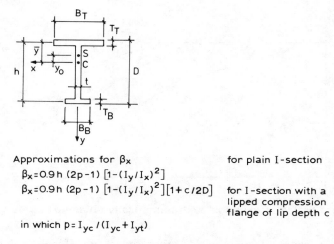

Approximations for β_x

$\beta_x = 0.9h(2p-1)\left[1-(I_y/I_x)^2\right]$ for plain I-section

$\beta_x = 0.9h(2p-1)\left[1-(I_y/I_x)^2\right][1+c/2D]$ for I-section with a lipped compression flange of lip depth c

in which $p = I_{yc}/(I_{yc}+I_{yt})$

Fig. 18. Properties of monosymmetric I-beams.

Figure 18 illustrates the determination of the more unusual of these properties, as well as providing good approximate expressions for β_x.[27] The principal effect of monosymmetry is to produce an increase in stability when the beam's larger flange is in compression and a decrease when the smaller flange is in compression as compared with a geometrical section having similar properties as illustrated in Fig. 19.

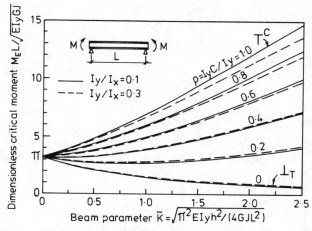

Fig. 19. Approximate critical moments of monosymmetric I-beams (after Ref. 27).

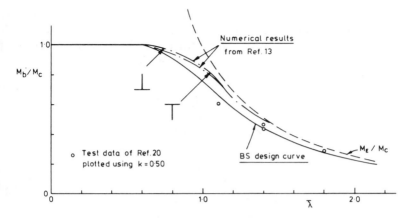

Fig. 20. Comparison of BS design curve with test data of Ref. 20 for unequal flanged I-beams.

On the basis of the test data of Hill[20] for beams with a moderate degree of monosymmetry ($I_c/I_y = 0.339$), Frey[9] has suggested the use of the same ECCS design curve with $\bar{\lambda}$ determined using M_E from eqn (5).

Figure 20 shows that the BS curve of Fig. 2 also provides very reasonable predictions of these four test results. Once again there is some uncertainty about the exact degree of end fixity present in the tests of Ref. 4; in Fig. 20 the maximum amount has been assumed with the effective length factor being taken as 0·5. In addition numerical ultimate strength results obtained by Lai[13] for a tee section with the flange in either tension or compression ($I_c/I_y = 0$ or 1) plot above the BS design curve over the whole range of $\bar{\lambda}$ as shown in Fig. 20.

6 TREATMENT OF SLENDER CROSS-SECTIONS

The design curves of Fig. 2 assume that at low values of $\bar{\lambda}$ beams will be able to sustain the full cross-sectional moment capacity M_c. It is well known, and generally recognised in recently produced codes of practice for metal construction,[5,6] that the particular value of M_c attainable will depend upon the slenderness of the individual plate elements of the cross-section. For beams Ref. 7 recognises three classes:

$$\begin{aligned} \beta &\leq \beta_1 & &\text{fully compact} \\ \beta_1 &\leq \beta \leq \beta_0 & &\text{semi compact} \\ \beta &> \beta_0 & &\text{slender} \end{aligned} \quad (6)$$

in which $\beta = f(b/t)$ with the exact relationship depending on the plate's edge support details and the pattern of stress to which it is subjected, and $b/t =$ ratio of flat width to thickness and the class of cross-section is that of its least favourable element.

The corresponding values of M_c will be:

$$\begin{aligned} M_c &= p_o S_n & \text{fully-compact} \\ M_c &= p_o Z_n & \text{semi-compact} \\ M_c &= p_o Z_e & \text{slender} \end{aligned} \qquad (7)$$

in which

$S_n =$ plastic section modulus,

$Z_n =$ elastic section modulus,

$Z_e =$ elastic modulus of effective section.

In determining Z_e an effective thickness $k_L t$, with the value of k_L being based on plate buckling data, is used for each plate element for which $\beta > \beta_o$.

For a semi-compact cross-section Ref. 7, in common with Ref. 5, neglects any possible deleterious interaction between local and lateral-torsional buckling, simply imposing a cut-off at a moment of $p_o Z_n$ as shown in Fig. 21.

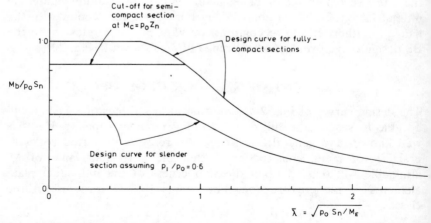

Fig. 21. Modifications to design curve for semi-compact and slender sections according to Ref. 7.

For slender sections various approaches to the local/lateral-torsional interaction are possible.[28] That used in Ref. 7 is based on studies originally conducted for light-gauge steel beams[29] and uses the idea of replacing the actual material strength p_o in the Perry-type of equation describing the design curve of Fig. 2 by a reduced value p_1 given by

$$p_1 = M_c/S_n \tag{8}$$

For simplicity $\bar{\lambda}$ is based on the properties of the gross cross-section. This avoids the need to calculate M_E for effective cross-sections that are monosymmetric due to the reduced effective thickness of the compression flange. A sample design curve assuming $p_1/p_0 = 0.6$ is shown in Fig. 21.

Referring back to the test data of Table 2, the cross-section used in all of the tests of Clark & Jombock[19] is semi-compact according to Ref. 7, since the flange b/T of 5·09 exceeds the compact limit of 4·68, whilst meeting the semi-compact limit of 5·49. Figure 22 replots these data, assuming that end restraint provides an effective length factor of 0·55, against the revised semi-compact lateral-torsional buckling design curve of Fig. 21. Compared with Fig. 11 only the test results for the two lowest values of $\bar{\lambda}$ are affected, the seemingly rather low result for $\bar{\lambda} = 0.69$ now falling only very slightly below the design curve, whilst the result for $\bar{\lambda} = 0.39$ (for which a moment in excess of $p_o S_n$ was obtained) now plots significantly above the curve.

Applying the same sort of correction to the results for other end moment ratios shown in Figs 13–16 (a horizontal cut-off at M_b/M_c of 0·84) means that all but one of the points at low values of $\bar{\lambda}$ that plot below the scaled up design curve (based on the direct use of the m

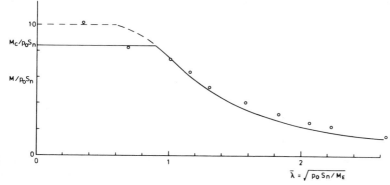

Fig. 22. Comparison of test results from Ref. 19 for $\beta = 1.0$ with design curve for semi-compact beams.

Table 4. Sections Tested by Cherry[31]

Section ref.	B (mm)	T (mm)	D (mm)	t (mm)	$\varepsilon = \sqrt{250/p_o}$	k_L	β/β_0	p_1/p_o
P	38·20	1·07	47·32	2·77	0·778	0·330	2·55	0·50
A	76·58	1·60	72·26	3·68	0·750	0·141	3·91	0·43
B	76·58	2·11	74·04	3·78	0·805	0·287	2·74	0·48
C	76·33	1·89	72·19	5·13	0·975	0·393	2·36	0·57
D	75·90	1·90	71·53	4·65	0·820	0·268	2·84	0·50

factor as shown) now plot above the revised design condition. However, since the tests on very stocky beams gave moments well in excess of $p_o S_n$ in several cases, limiting the design to $p_o Z_n$ would appear to be unduly conservative at this end of the range. The only available set of test data for the lateral-torsional buckling of slender aluminium beams is that produced by Cherry[30] for sections with very thin flanges. For the five I-sections Table 4 provides details. This shows that, according to Ref. 7, not only are the sections slender but the flange proportions are such that very low values of k_L (Ref. 28) are appropriate, leading to values of Z_e significantly below Z_n and thus values of p_1 significantly below p_o. Because Ref. 30 gives only 0·1% proof stress values for the materials used, ε, the material strength correction on the β limits, has been determined taking p_o as $\sigma_{0·1}$.

A comparison of 35 test results from Ref. 30 and the design procedure of Ref. 7 is given in Fig. 23. This shows the procedure to be generally acceptable—a few results for the series A beams with extremely slender flanges plot below the design curve—if somewhat conservative at low beam slenderness. Thus actual local/overall interaction would appear to be satisfactorily handled by the method of Ref. 7. The results of Fig. 23 do, however, suggest that the allowances made therein for loss of effectiveness of very slender outstands may be too severe.

7 CONCLUSIONS

The basis for the design of laterally unrestrained aluminium beams of open cross-section has been reviewed. Test data and ultimate strength numerical results have been used to quantify the effects of key parameters on behaviour. It has been shown that design procedures,

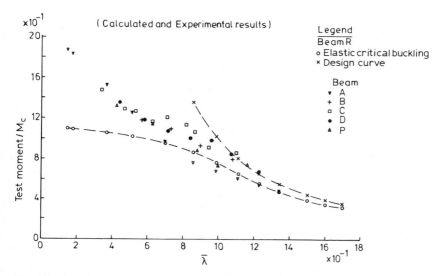

Fig. 23. Comparison of design procedure for lateral-torsional buckling of slender sections with test data of Ref. 30.

including beam design curves, originally devised for steel members may be used, providing final 'tuning' of such procedures is conducted using data that relate specifically to aluminium.

ACKNOWLEDGEMENTS

In conducting the work on which this chapter is based the author wishes to acknowledge the original contributions of Dr Y. W. F. Lai and the useful discussions with members of the BSI CSB 36 committee, especially Dr P. S. Bulson and Mr J. B. Dwight.

REFERENCES

1. Nethercot, D. A., Elastic lateral buckling of beams. In *Beams and Beam Columns: Stability and Strength*, Chapter 1, ed. R. Narayanan. Applied Science Publishers, London, 1983, pp. 1–34.
2. ECCS, European Recommendations for Steel Construction. European Convention for Constructional Steelwork, 1978.
3. ECCS, European Recommendations for Aluminium Alloy Structures. European Convention for Constructional Steelwork, 1978.

4. Trahair, N. S., In discussion of Lateral stability of steel I-beams in the plastic range, by C. Massey. *Civil Engineering Transactions*, Institution of Engineers, Australia, Vol. CE 6, No. 2. September 1964, pp. 119–29.
5. British Standards Institution, *BS 5950: Structural Use of Steelwork in Building: Part 1 Code of Practice for Design in Simple and Continuous Construction: Hot-Rolled Sections*. BSI, London, 1990.
6. British Standards Institution, *BS 5400: Steel, Concrete & Composite Bridges: Part 3 Code of Practice for the Design of Steel Bridges*. BSI, London, 1982.
7. British Standards Institution, *BS 8118: The Structural Use of Aluminium*. BSI, London, 1992.
8. Dwight, J. B., Use of the Perry formula to represent the European column curves. *Steel Construction*, Vol. 9, No. 1, Australian Institute of Steel Construction, 1975.
9. Frey, F., Buckling, lateral buckling and eccentric buckling of alu-alloy columns, beams and beam-columns. ECCS Committee 16, Doc. 16-77-1, March 1977.
10. Structural Stability Research Council, *Guide to Stability Design Criteria for Metal Structures*, 4th edn ed T. V. Galambos. Wiley Interscience, New York, 1988.
11. Mazzolani, F. M., *Aluminium Alloy Structures*. Pitman, London, 1985.
12. Hong, G. M., Aluminium column curves. In *Aluminium Structures, Advances, Design and Construction*, ed. R. Narayanan. Elsevier Applied Science, London, 1987, pp. 40–9.
13. Lai, Y. F. W., Buckling strength of welded & non-welded aluminium beams. PhD thesis, University of Sheffield, 1988.
14. Nethercot, D. A. & Trahair, N. S., Design rules for the lateral buckling of steel beams. CE19, No. 2, *Civil Engineering Transactions*, Institution of Engineers, Australia, 1977, pp. 162–5.
15. Kloppel, K. & Barsch, W., Versucke zum Kapitel Stabilitatstalle Der Nevfassung von DIN 4113. *Aluminium*, Helft 10 (1973) 690 (English translation available from MOD).
16. Dumont, C. & Hill, H. N., The lateral instability of deep rectangular beams. NACA Tech. Note No. 601, 1937.
17. Clark, J. W. & Rolf, R. L., Buckling of columns, plates and beams. *Journal of the Structural Division, ASCE*, **No. ST3** (1966) 17–38.
18. Dumont, C. & Hill, H. N., The lateral stability of equal-flanged alu-alloy I-beams in pure bending. NACA Tech. Note No. 770, 1940.
19. Clark, J. W. & Jombock, J. R., Lateral buckling of I-beams subjected to unequal end moments. *Journal of the Engineering Mechanics Division, ASCE*, **EM3** (1957).
20. Hill, H. N., The lateral instability of unsymmetrical I-beams. *Journal of the Aeronautical Sciences*, **9** (1942) 175–179.
21. El-Khenfas, M. A., Analysis of beam-column strength under biaxial loading. PhD thesis, University of Sheffield, 1987.
22. Wang, Y. C., Biaxial bending of end-restrained columns. PhD thesis, University of Sheffield, 1988.
23. Lindner, J. L. & Gietzelt, R., Discussion of interaction equations for

members in compression and bending. 3rd International Colloquium on Stability of Metal Structures, CITIM, Paris, 1983, Final Report, pp. 45–54.
24. El-Khenfas, M. A., Wang, Y. C. & Nethercot, D. A., Lateral torsional buckling of end-restrained beams. *Journal of Constructional Steel Research*, **7**(3) (1987) 335–63.
25. Nethercot, D. A., Effective lengths of partially plastic steel beams. *Journal of the Structural Division, ASCE*, **101** (ST5) (May 1975) 1163–6.
26. Wood, R. H., *A New Approach to Column Design*. HMSO, London, 1974.
27. Kitipornchai, S. & Trahair, N. S., Buckling properties of monosymmetric I-beams. *Journal of the Structural Division, ASCE*, **106** (ST5) (May 1980) 941–57.
28. Kubo, M. & Fukumoto, Y., Lateral torsional buckling of thin-walled I-beams. *Journal of the Structural Division, ASCE*, **114** (ST4) (April 1988) 841–55.
29. Davies, J. M. & Thomanon, P. O., Local and overall buckling of light gauge members. In *Instability and Plastic Collapse of Steel Structures*, ed. L. J. Morris, Granada, 1983, pp. 479–92.
30. Cherry, S., The stability of beams with buckled compression flanges. *The Structural Engineer* (Sept. 1960) 277-85.

4

Shear Webs and Plate Girders

H. R. EVANS
University of Wales College of Cardiff, UK

ABSTRACT

This chapter describes in detail the procedures specified in the new design code BS 8118 for determining the resistance of plate girders to various loads. The detailed equations in the code are presented and explained and, in addition, the general philosophy of the procedures is outlined, together with the background theory. Detailed comparisons between values determined by the new code and those of the previous code (CP 118) are summarised and show the new code to predict higher girder capacities in most cases. A summarised comparison of the new code values and recently obtained experimental data is also presented and shows the code to provide safe, although rather conservative, predictions of strength.

It should be noted that this chapter has as its reference the 1985 draft of BS 8118, which was widely distributed. The final version of BS 8118 (1992) has a revised nomenclature and a new format for the design rules. The basic procedures and detailed comparisons are, however, unaltered. Clause numbers quoted in the chapter refer to the final version of BS8118.

1 INTRODUCTION

A fabricated or built-up plate girder, such as that shown diagrammatically in Fig. 1 is generally required to support vertical loads over long spans and has to resist the high bending moments and shearing forces resulting therefrom. The primary function of the top and bottom

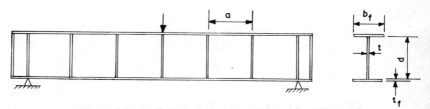

Fig. 1. Typical simply supported plate girder showing notation used in BS 8118 for girder dimensions.

flanges is to resist the axial compressive and tensile forces arising from the bending action whilst the web plate resists the shear force. Connections, whether welded, riveted or bolted, between the web and flange plates must ensure the adequate transference of longitudinal shear.

For a given applied bending moment, the axial forces in the flanges decrease as the web depth (d) is increased so that, from this point of view, it is advantageous to make the webs as deep as possible. To reduce the self weight of the girder, the web thickness (t) is usually limited, with the consequence that the webs are normally of thin-walled proportions (web proportions are normally expressed in terms of the slenderness ratio $\dfrac{d}{t}\sqrt{\dfrac{f_{0\cdot 2}}{250}}$, where $f_{0\cdot 2}$ is the 0·2% proof strength of the material). The webs will then buckle at relatively low values of the applied shear loading.

In order to provide an efficient and economical girder design, advantage should be taken of the post-buckling capacity of the girder, i.e. its ability to carry loads considerably in excess of that at which the web buckles. In this way, a girder of high strength/weight ratio can be designed, suitable for use in situations where reduction of self-weight is of prime importance. Aluminium, of course, is normally only used in such situations.

The post-buckling strength of transversely stiffened plate girders is taken into full account in the recently published design codes for steel bridges (BS 5400)[1] and buildings (5950)[2] and a considerable amount of experimental and theoretical information[3] is available on the behaviour of *steel* girders in the post-buckling range. Because of the significantly reduced material stiffness, resulting in earlier buckling, it is even more important to allow for post-buckling capacity in aluminium girders than in steel. However, there is much less experimental information regarding the post-buckling behaviour of aluminium girders. Consequently, the procedures adopted by the new Code of Practice for the Structural Use of Aluminium (BS 8118) have been based upon those originally developed for steel girders. There is, however, one important modification for *welded* aluminium girders to allow for the reduction in strength within the heat-affected zone (HAZ) adjacent to a weld. HAZ softening has been discussed in Chapter 1 and appropriate values of the associated strength reduction factor w will be assumed in this chapter.

Tests reported recently on 22 welded girders of 6082 alloy[4] and on a further six welded girders of the same alloy[5] have justified the procedures of the new code, whilst indicating a considerable degree of conservatism. These experimental studies will be summarised later in this chapter and the girder capacities predicted by the code will be compared to those measured experimentally. A comparison of the capacities given by the new code to those of CP 118[6] will also be included.

2 GENERAL OBSERVATIONS ON THE NEW CODE

As discussed in earlier chapters, the basic design requirement that the new code sets for plate girders is that the factored resistance should exceed the action produced by the factored design loading. This chapter will concentrate upon the procedures adopted for the determination of the factored resistance.

Three different types of plate girders are identified by the code: unstiffened girders (Clause 4.5.3), transversely stiffened girders (Clauses 5.4.2 and 5.4.3) and girders with longitudinal and transverse stiffeners (Clause 5.4.4). Since there is no experimental evidence available for the post-buckling capacity of unstiffened aluminium girders, their capacity is limited to that load at which buckling will first occur; a similar approach is adopted in the new steel buildings code (BS 5950). Transversely stiffened girders are designed to their full post-buckling capacity using tension field procedures similar to those adopted by all the new steel codes, although these procedures have been considerably simplified in the aluminium code; the influence of material softening within the HAZ is also taken into account. Longitudinally stiffened girders are also designed using a similar tension field approach; this approach has not been adopted in the steel bridge code (BS 5400) although, on the basis of considerable experimental evidence, it has recently been proposed for steel girders by the ECCS Technical Working Group 8.3—Plated Structures.[7]

To ensure that girders develop their full capacity, the web stiffeners must be carefully designed. The design of such stiffening elements is detailed in Clause 5.4.5.

In many cases during the design of buildings, openings have to be cut in girder web plates to allow the passage of service ducts. There is no available experimental evidence on the behaviour of aluminium girders with web openings; consequently, the capacity of any web containing a large opening is limited to that load at which buckling will first occur on the assumption that there is no tension field capability.

In all clauses to this stage, the code concentrates upon the determination of girder capacity under pure shear. The effects of a co-existing bending moment upon shear capacity is then considered in Clause 5.4.7, adopting procedures similar, though not identical, to those of the steel codes.

Before considering individual girder types, it is possible to identify one common aspect in that, for all classes of girder, the factored shear resistance (V_{RS}) is related to the basic shear capacity (V_1) of the web as follows:

$$V_{RS} = \frac{V_1}{\gamma_m} \times \text{reduction factor } (C_S \text{ or } C_{ST}) \text{ where } C_S \text{ and } C_{ST} < 1. \quad (1)$$

where

γ_m is the partial safety factor on material strength, taken as 1·2 for unwelded or welded webs,

V_1 is the basic shear capacity obtained from the material design stress f_d (normally taken as the 0·2% proof stress $f_{0·2}$), suitably reduced, where necessary, by the factor w to allow for HAZ softening in welded girders, as follows:

$$V_1 = w \, dt \frac{f_d}{\sqrt{3}} = w \, dt \frac{f_{0·2}}{\sqrt{3}} \text{ (normally)} \quad (2)$$

The reduction factor in eqn (1) is introduced to allow for the fact that the onset of buckling will prevent the full value of the basic shear capacity (which is equivalent to the full yield capacity of a steel girder) from being attained in a slender web. The reduction factor is termed

C_S for unstiffened webs and C_{ST} for stiffened webs and the calculation of appropriate values of these factors will now be considered in some detail. However, it is worth emphasising at this stage that neither C_S nor C_{ST} should ever be allowed to exceed unity, as would be expected for *reduction* factors.

The definition of web slenderness ratio is also common to both unstiffened and transversely stiffened girders; in general terms:

$$\text{slenderness ratio} = \frac{d}{t}\sqrt{\frac{f_{0.2}}{250}} \qquad (3)$$

In the latest version of BS 8118, the use of f_d, w, C_S and C_{ST} has been discontinued, and an alternative nomenclature and analytical method introduced. This, however, does not alter the fundamental calculation process or design philosophy discussed here.

3 UNSTIFFENED WEBS IN SHEAR (CLAUSE 5.6.2)

Any web plate that is not reinforced by transverse stiffeners positioned at a spacing (a) equal to, or less than, 2·5 times the web depth (d) is regarded as being 'unstiffened'. Thus, for an unstiffened web:

$$a/d > 2\cdot 5$$

3.1 Webs free from HAZ softening

Since no experimental studies have been carried out of the post-buckling capacities of unstiffened aluminium webs, their shear capacity is limited to that load at which buckling will first occur; a similar approach was followed in the previous code (CP 118). The new steel buildings code (BS 5950) imposes a similar restriction although, in the case of unstiffened steel girders, there is some experimental evidence[8] of post-buckling action.

The critical shear stress of a web plate of width a and depth d can be determined from elastic buckling theory.[9] It is a function of the boundary conditions of the plate, but the true boundary conditions for a girder web are difficult to establish accurately because the degree of restraint imposed by the flanges and by the adjacent web panels cannot be evaluated. However, it can be assumed conservatively that all the boundaries of the web panel are simply supported, so that the critical shear stress (τ_{cr}) is then given by:

$$\tau_{cr} = k \frac{\pi^2 E}{12(1-v^2)} \left(\frac{t}{d}\right)^2 \tag{4}$$

where the buckling coefficient k is obtained as

$$k = 5\cdot 35 + 4\left(\frac{d}{a}\right)^2 \quad \text{when} \quad \frac{a}{d} \geq 1$$

and

$$k = 5\cdot 35\left(\frac{d}{a}\right)^2 + 4 \quad \text{when} \quad \frac{a}{d} \leq 1$$

Assuming typical values of elastic properties for aluminium, i.e. $E = 70\,000\,\text{N/mm}^2$, $v = 0\cdot 34$, the buckling stress for a panel with $a/d \geq 1$ becomes:

$$\tau_{cr} = \frac{348\,274[1 + 0\cdot 75(d/a)^2]}{(d/t)^2}$$

The ratio between the corresponding buckling load (V_{cr}) and the basic shear capacity ($V_1 = dt f_{0\cdot 2}/\sqrt{3}$ for an unwelded web from eqn (2)) may be expressed by introducing a new term v_1 (which will reappear later for stiffened webs), where:

$$v_1 = \frac{V_{cr}}{V_1} = \frac{348\,274[1 + 0\cdot 75(d/a)^2]}{(d/t)^2(f_{0\cdot 2}/\sqrt{3})}$$

To express v_1 in terms of the slenderness ratio defined in eqn (3), the numerator and denominator of this expression may be multiplied by $f_{0\cdot 2}/250$ to give:

$$v_1 = \frac{2400[1 + 0\cdot 75(d/a)^2]}{\left(\dfrac{d}{t}\sqrt{\dfrac{f_{0\cdot 2}}{250}}\right)^2} \quad \text{for} \quad \frac{a}{d} \geq 1 \tag{5}$$

and similarly

$$v_1 = \frac{2400[(d/a)^2 + 0\cdot 75]}{\left(\dfrac{d}{t}\sqrt{\dfrac{f_{0\cdot 2}}{250}}\right)^2} \quad \text{for} \quad \frac{a}{d} \leq 1$$

For the particular case of an unstiffened girder, the panel width (a)

is very large compared to the web depth (d) so that the term (d/a) becomes negligible. Also, since the capacity of an unstiffened girder, i.e. $C_S V_1$ from eqn (1), is to be limited to its buckling capacity (V_{cr}), the reduction factor (C_S) becomes equal to v_1 in this particular case. Thus:

$$C_S = v_1 \approx 2400 \bigg/ \left(\frac{d}{t}\sqrt{\frac{f_{0\cdot 2}}{250}}\right)^2 \not> 1\cdot 0 \qquad (6)$$

this being the form in which the equation appeared in the 1985 draft code.

From eqn (6) it can be seen that buckling will not occur, i.e. $C_S = 1$, when the web slenderness becomes:

$$\frac{d}{t}\sqrt{\frac{f_{0\cdot 2}}{250}} \leq 49 \qquad (7)$$

Less slender webs will then achieve the full value of the basic shear capacity V_1.

3.2 Unstiffened webs with HAZ softening

Even in an unstiffened web, certain attachments, for example transverse stiffeners at a spacing greater than $2\cdot 5d$, load bearing stiffeners or end posts, will be welded to the girder. The welds to attach these elements, and the corresponding HAZs, will extend over the complete web depth so that, as in eqn (2), the basic shear capacity will be reduced by the introduction of the softening factor w, i.e.

$$V_1 = w \, dt \frac{f_{0\cdot 2}}{\sqrt{3}}$$

Such welds, and those connecting the web to the flanges, and the associated HAZs will be limited to the perimeter of the web plate. Consequently, they will have little effect upon the buckling capacity and it may be assumed that the buckling capacity of a welded unstiffened web is identical to that of the corresponding unwelded web. However, to allow for the reduced value of the basic shear capacity, V_1, as above, eqn (6) must be written as:

$$C_S = \frac{V_1}{\omega} \approx \frac{2400}{w\left(\frac{d}{t}\sqrt{\frac{f_{0\cdot 2}}{250}}\right)^2} \qquad (8)$$

From this equation it can be seen that buckling will not occur, i.e. $C_S = 1$, when the web slenderness becomes:

$$\frac{d}{t}\sqrt{\frac{f_{0\cdot 2}}{250}} \leq \frac{49}{\sqrt{w}} \qquad (9)$$

4 TRANSVERSELY STIFFENED WEBS IN SHEAR (CLAUSE 5.6.3)

A web is classified as 'transversely stiffened' when it is reinforced by stiffeners positioned at a spacing (a) not greater than 2·5 times the web depth (d). A stiffener spacing of less than $0\cdot 5d$ is not considered since the resulting design would be uneconomical. Thus, for a transversely stiffened web:

$$0\cdot 5 \leq \frac{a}{d} \leq 2\cdot 5$$

and there is little experimental evidence of the performance of aluminium girders outside this range.

4.1 Webs—free from HAZ softening

Transversely stiffened girders in shear are designed to their full post-buckling capacity, using tension field procedures similar to those adopted by the new steel codes. It has been well-documented elsewhere[3] that a transversely stiffened girder can withstand loads considerably in excess of the load at which web buckling first occurs. This post-buckling reserve of strength arises from the development of tension field action within the web, as illustrated diagrammatically in Fig. 2(a).

Once a web has buckled, it loses its capacity to carry additional compressive stresses so that, in the post-buckling range, a new load-carrying mechanism is developed, whereby any additional shear load is carried by an inclined tensile membrane stress field. This tensile field anchors against the top and bottom flanges and against the transverse stiffeners, as shown. The load carrying action of the girder then becomes similar to that of the Pratt truss in Fig. 2(b), with the resistance offered by the web plates being analogous to that of the diagonal tie bars in the truss; the view of the typical steel girder after collapse in Fig. 2(c) confirms this behaviour.

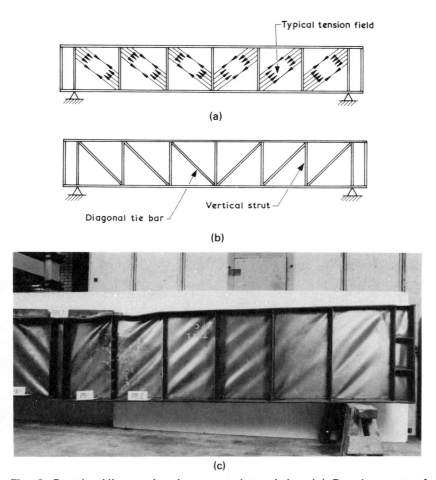

Fig. 2. Post-buckling action in a complete girder. (a) Development of tension field action in adjacent panels. (b) Pratt truss analogy. (c) Development of tension field action in a real girder.

The girder collapses when the web yields and four plastic hinges form in the flanges to allow the formation of a shear sway collapse mechanism. This assumed mechanism is shown diagrammatically in Fig. 3(a) and confirmed by the view of a typical steel girder after collapse in Fig. 4(a). Figure 4(b) gives a similar view of an aluminium girder of 6082 alloy which was tested as part of a detailed experimental study reported in Ref. 4.

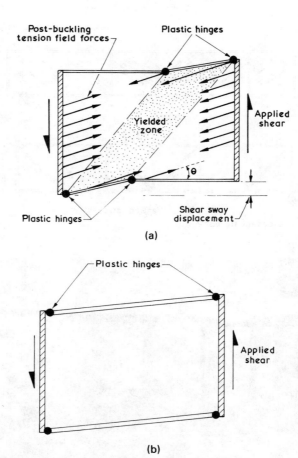

Fig. 3. Assumed collapse mechanism in an individual panel. (a) Assumed shear sway mechanism in a typical panel. (b) 'Picture frame' mechanism in a panel with very strong flanges.

An analysis of the collapse mechanism, as detailed in Ref. 3, results in the following general expression for the shear capacity (V_S), non-dimensionalised with respect to the basic shear capacity (V_1) of an *unwelded* web:

$$\frac{V_s}{V_1} = v_1 + \sqrt{3}\sin^2\theta\left(\cot\theta - \frac{a}{d}\right)\frac{f_t^y}{f_{0\cdot 2}} + \sqrt{\frac{M_{pf}}{M_{pw}}} \times 2\sqrt{3}\sin\theta\sqrt{\frac{f_t^y}{f_{0\cdot 2}}} \quad (10)$$

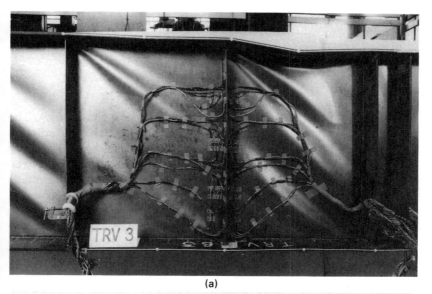

(a)

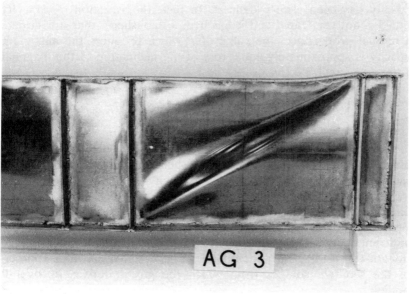

(b)

Fig. 4. Views of girders after collapse. (a) Typical steel girder. (b) Typical aluminium girder.

For convenience, this may be written as:

$$\frac{V_s}{V_1} = v_1 + v_2 + m^* v_3 = C_{ST} \ngtr 1 \cdot 0 \tag{11}$$

where $V_1 = dt(f_{0 \cdot 2}/\sqrt{3})$ from eqn (2).

Each of these three terms has a particular physical significance. The first term (v_1) represents the buckling capacity, as derived earlier in eqn (5). The second term (v_2) represents the strength derived from that part of the tension field supported by the transverse web stiffeners, and the third term (m^*v_3) represents the contribution made by the flanges to the overall girder capacity.

In the expression given in eqn (10), the term θ represents the inclination of the web tension field, see Fig. 3(a). The value of θ cannot be determined directly and, in a full analysis, an iterative procedure has to be adopted in which successive values of θ are assumed and the corresponding ultimate load evaluated in each case. These iterations have to be repeated until the maximum value of shear capacity has been clearly defined. In fact, the variation of capacity with θ is not very rapid and it has been established[10] that, for girders of normal proportions, the value of θ which produces the maximum capacity is approximately equal to $\tfrac{2}{3}$ of the inclination of the diagonal of the web panel, i.e.

$$\theta = \frac{2}{3} \tan^{-1}\left(\frac{d}{a}\right) \tag{12}$$

The assumption of such a value of θ will lead either to the correct value or to an underestimation of the shear capacity; it is thus a safe assumption to make in the context of design.

This adoption of an explicit value for θ has a significant effect in simplifying the design procedure. It allows the second and third terms of the expression in eqn (11) to be considered independently so that the component of girder capacity that is dependent upon flange strength (m^*V_3) is clearly identifiable. The designer can thus appreciate the separate sources from which the girder derives its overall strength.

The derivation of each of the three terms of eqn (11) in a form suitable for design calculations will now be considered. The first term (v_1) has already been derived in eqn (5) and its variation with the web slenderness parameter $\dfrac{d}{t}\sqrt{\dfrac{f_{0\cdot 2}}{250}}$ can be conveniently represented by a family of curves, as in Fig. 5(a). Each curve corresponds to a different web aspect ratio (a/d), upon which the buckling coefficient is dependent. At low values of web slenderness, each curve reaches the value of 1 indicating that the full basic shear capacity of the web would be achieved. In such cases, of course, there would be no post-buckling action so that the second and third terms (v_2 and m^*v_3) would be zero.

The second term (v_2) represents the post-buckling strength derived from the web tension field anchoring on the transverse stiffeners. From eqn (10), the explicit form of this term is given by:

$$v_2 = \sqrt{3}\sin^2\theta\left(\cot\theta - \dfrac{a}{d}\right)\dfrac{f_t^y}{f_{0\cdot 2}} \tag{13}$$

It is thus a function of the tension field stress (f_t^y) developed in the web at yield in the case of steel girders, or at the 0·2% proof stress ($f_{0\cdot 2}$) in the case of aluminium girders. This has been determined (see Ref. 3) from the von Mises–Hencky yield criterion as:

$$\dfrac{f_t^y}{f_{0\cdot 2}} = \sqrt{1 - v_1^2\left(1 - \dfrac{3}{4}\sin^2 2\theta\right)} - \dfrac{\sqrt{3}}{2}v_1\sin 2\theta \tag{14}$$

The tension field stress thus becomes a function of v_1 which, as we have seen in eqn (5) above, is dependent on the web slenderness parameter. The second term (v_2) of eqn (11) thus becomes a function of web slenderness and the rather complex expressions in eqns (13) and (14) may again be presented as a family of convenient design curves showing the variation of v_2 with slenderness for different values of web panel aspect ratio (a/d); these curves are given in Fig. 5(b).

The third term (m^*v_3) in eqn (11) represents the component of post-buckling capacity that is dependent on the strength of the flanges. It encompasses and is dependent upon that part of the web tension field that anchors against the flanges (v_3), as well as the plastic moment capacity of the flange plates themselves (m^*). From eqn (10), the explicit form of v_3 and m^* may be written as:

$$v_3 = 2\sqrt{3}\sin\theta\sqrt{\dfrac{f_t^y}{f_{0\cdot 2}}}$$

$$m^* = \sqrt{\dfrac{M_{pf}}{M_{pw}}} \not> m_1 \tag{15}$$

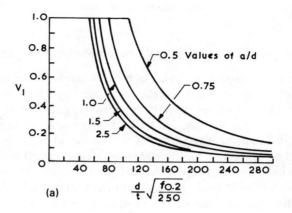
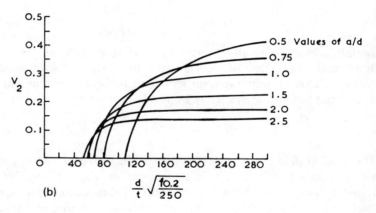
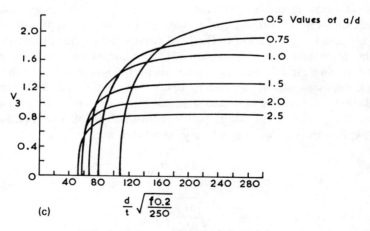

Fig. 5. Values of (a) v_1; (b) v_2; (c) v_3.

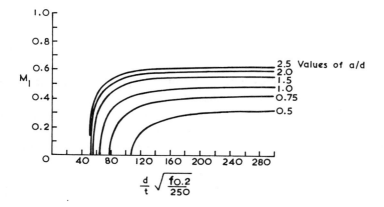

Fig. 5.—*contd.* (d) M_1.

Thus, v_3 is again seen to be a function of the tension field stress (f_t^y) and can, therefore, again be expressed in terms of the web slenderness parameter. The family of design curves defining this relationship is given in Fig. 5(c).

On the other hand, m^* is simply a non-dimensional parameter relating the plastic moment capacity of the flange ($M_{pf} = S_f f_{0.2f}$ for a flange plate having a plastic modulus S_f about its own equal area axis and a proof stress $f_{0.2f}$) to that of the web ($M_{pw} = 0.25td^2 f_{0.2}$). When the proof stresses of the web and flange material are the same, then:

$$m^* = \sqrt{\frac{4S_f}{td^2}} \not> m_1 \qquad (16)$$

In the particular case of a flange of rectangular cross-section of width b_f and thickness t_f, $S_f = 0.25 b_f t_f^2$. An upper limit of m_1 is imposed upon the value of m^* because the position of the internal plastic hinge forming in each flange at collapse, see Fig. 3(a), is a function of flange strength. As the flange strength increases, the internal hinge approaches the extreme end of the panel, finally reaching the end of the panel for strong flanges, to form a 'picture frame' collapse mechanism, as shown in Fig. 3(b). This represents a limiting case of the assumed collapse mechanism and, for simplicity, the value of m^* is limited to that at which this condition is reached;

this limiting value is defined by:

$$m_1 = \frac{a}{d}\sin\theta\sqrt{\frac{f_t^y}{f_{0\cdot 2}}} \qquad (17)$$

Thus, m_1 again becomes a function of web slenderness and can be represented by the family of design curves in Fig. 5(d). It should, however, be stressed that this limit would not be approached by girders of normal proportions so that the rather conservative approach adopted here for simplicity would not lead to undue conservatism in most cases.

The four sets of curves given in Fig. 5(a)–(d) enable the shear capacity of a transversely stiffened girder to be determined for any web slenderness and for any web aspect ratio within the range $0\cdot 5 \leq a/d \leq 2\cdot 5$. There is little available experimental evidence of the performance of aluminium girders with aspect ratios lying outside this range. The curves are plotted for web slenderness ratios up to 300. This maximum value represents an extremely slender web with, as shown in Fig. 5(a), a very low buckling capacity. Such webs do, however, possess a significant post-buckling reserve of strength. This is well-illustrated in Fig. 6, where the variation of each of the three terms in eqn (10) with web slenderness is plotted for typical transversely stiffened girders. It is clear that, at high web slendernesses, the strength is almost entirely derived from post-buckling action, with the flange dependent contribution being significant.

Such slender webs will, of course, be heavily buckled prior to collapse, as indicated by the photograph in Fig. 4(b), and will possibly show signs of buckling at working loads. This will obviously have implications in terms of serviceability and fatigue which the designer must consider and balance against the potential of very slender webs to provide a significant saving of self weight.

4.2 Transversely stiffened webs with HAZ softening

In contrast to the fully ductile failures observed in steel girders, many welded aluminium girders fractured at collapse. A typical fracture is apparent in Fig. 4(b), occurring within the softened material in the heat affected zone adjacent to the flange/web and stiffener/web welds in the top right hand corner of the panel. Such fractures are illustrated and described in more detail in Ref. 4, which reports on an extensive

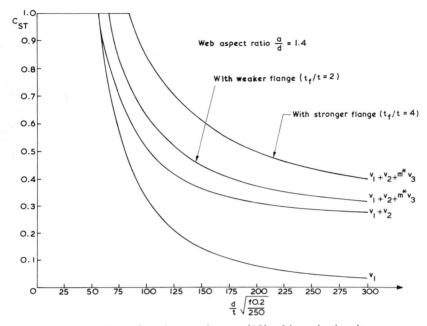

Fig. 6. Variation of each term in eqn (10) with web slenderness.

experimental study of the collapse behaviour of aluminium girders of 6082 alloy.

These fractures invariably occurred at a late stage of the experiments when the large deformations associated with the formation of a shear sway mechanism were developing and when the load/deflection curve was already indicating the start of a 'plastic' failure plateau (it should be appreciated that such a plateau is not expected to be well-defined in the case of a material such as aluminium which has no distinct yield point). The fractures may, therefore, be regarded as the consequence, rather than the cause, of failure.

However, since a plastic method of design is being used in the Code, implicitly assuming ductile material behaviour, it is important to guard against the possibility of such 'brittle' failures.

Currently, there is insufficient experimental evidence on the collapse of welded aluminium girders to allow an accurate collapse mechanism, incorporating the possibility of web fracture in the heat affected zone, to be established. Thus, to retain simplicity, a very conservative approach is adopted in the Code wherein it is assumed that the whole area of the web plate suffers a reduction in strength as a result of welding; it is, of course, known that the HAZ is limited to a narrow band around the perimeter of the web, adjacent to the welds. There is, however, some justification for the approach adopted by the Code since, as indicated in Fig. 3(a), the tension field forces developed within the web have to pass through the softened perimeter regions before they can anchor against the flanges and transverse stiffeners; yield can, of course, occur locally within these perimeter regions.

The consequence of this simplifying assumption is that the whole of the post-buckling strength of the girder is reduced by the effects of welding. From eqn (11) it is known that the post-buckling capacity is given by $(v_2 + m^* v_3) V_1$ so that simply taking the reduced basic shear capacity of a welded web $(V_1 = w \, dt f_{0.2}/\sqrt{3})$ allows the required reduction to be achieved automatically. The calculation of v_2, m^* and v_3 then remains the same as for unwelded webs in Section 4.1. Although yielding can occur locally, buckling involves the complete area of the web plate so that, as discussed earlier in Section 3.2, softened zones around the plate perimeter will have little effect upon the buckling capacity. Then, as seen earlier in eqn (8), the buckling term v_1 must be divided by w to compensate for the reduction in V_1.

Thus, the shear capacity (V_S) of a welded web may be obtained by modifying eqn (11) as follows:

$$\frac{V_S}{V_1} = \frac{v_1}{w} + v_2 + m^* v_3 = C_{ST} \not> 1 \cdot 0 \tag{18}$$

where $V_1 = w \, dt \frac{f_{0.2}}{\sqrt{3}}$ from eqn (2), all other terms, i.e. v_1, v_2, v_3 and m^* being calculated as for unwelded webs in Section 4.1.

The curves already presented in Figs 5(a–d) can then be used again to determine the values of v_1, v_2, v_3 and m_1 and, consequently, the shear capacity of welded transversely stiffened webs. The resulting design procedure is, therefore, simple, although rather conservative.

5 LONGITUDINALLY STIFFENED WEBS IN SHEAR (CLAUSE 5.6.4)

Transversely stiffened webs may be further reinforced by the addition of longitudinal stiffeners, this could prove economical in certain circumstances.

On the basis of considerable experimental evidence it has been postulated,[3] and recently accepted[7] as a design procedure, that tension field theory can be applied to determine the post-buckling capacity of longitudinally stiffened steel girders. Indeed, longitudinally stiffened webs may be treated in virtually the same way as webs with transverse stiffeners only, on the assumption that the main effect of the longitudinal stiffeners is to increase the initial buckling resistance of the web. After buckling, it may be assumed that the tension field develops over the complete web depth, with the influence of the longitudinal stiffeners upon the post-buckling action being neglected.

Recently, to confirm the earlier experimental justification of this assumption in the case of steel girders,[3] similar experimental evidence has been gathered[4] for longitudinally stiffened aluminium girders; a photograph of a typical girder after failure is shown in Fig. 7(a). The development of an overall tension field and a shear sway failure mechanism, similar to those shown earlier in Fig. 4 for transversely stiffened girders, is clearly shown in the photograph.

On the basis of this assumed behaviour, a simple and conservative design procedure, similar to that described earlier for girders with transverse stiffeners only, may be adopted for longitudinally stiffened girders. The major difference arises in the calculation of the initial buckling load, where the depth of the largest web sub-panel (d' in Fig. 7(b)) is taken for longitudinally stiffened girder, since this sub-panel will buckle first, in place of the overall web depth.

Thus, as seen earlier in eqn (1), the factored shear resistance of a longitudinally stiffened girder is obtained as $V_R = V_1 C_{ST}/\gamma_m$, the reduction factor C_{ST} again being written from eqn (11) as:

$$C_{ST} = v_1 + v_2 + m^* v_3 \not> 1 \cdot 0 \qquad (19)$$

The first term, v_1, representing the buckling capacity of a longitudinally stiffened web, may be obtained from the family of curves already presented in Fig. 5(a). Since it is now being related to the the largest sub-panel, then the value of d' from Fig. 7(b) is taken in the calculation of the web slenderness parameter $\dfrac{d'}{t} \sqrt{\dfrac{f_{0 \cdot 2}}{250}}$ and web aspect ratio $\dfrac{a}{d'}$.

The two remaining terms are related to the post-buckling action that is assumed to develop over the complete web depth (d) and are, therefore, obtained directly from the equations derived earlier for

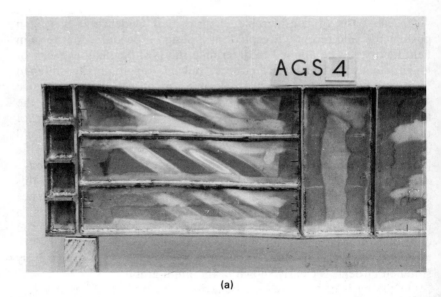

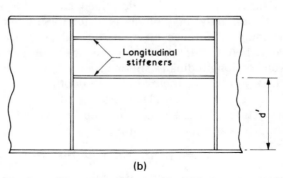

Fig. 7. Considerations for longitudinally stiffened girders. (a) Development of typical shear sway mechanism. (b) Definition of d' as depth of largest sub-panel.

transversely stiffened webs. It should be noted that, since each of these terms is a function of v_1, and since, as discussed above, the definition of v_1 has now changed, the design curves presented earlier in Figs 5(b–d) cannot be used for longitudinally stiffened webs. Similar curves could of course be provided, but a different set of curves would be required for each possible longitudinal stiffener position. Since it is not expected that longitudinally stiffened webs will be employed all that frequently, and since the basic equations are reasonably simple to use anyway, the inclusion of such curves in the draft code was not considered to be justified.

Thus, for a longitudinally stiffened web free from HAZ softening, having obtained the buckling capacity (v_1) from Fig. 5(a), as described above, the values of v_2 and m^*v_3 are obtained from eqns (13)–(17), taking the overall web depth (d) in the calculations.

For a longitudinally stiffened web that is welded, the effects of softening within the HAZ are taken into account by introducing the factor w, exactly as described earlier for transversely stiffened webs. Thus, eqn (18) may again be used to give a simple, conservative procedure.

6 REQUIREMENTS FOR STIFFENERS (CLAUSE 5.6.5)

To allow the girders to develop the capacities described above, careful attention must be paid to the design of the web stiffeners. As illustrated in Fig. 8, these may be of the following types:

intermediate transverse stiffeners,
transverse bearing stiffeners,
bearing stiffeners or end posts,
longitudinal stiffeners.

Each stiffener can be single or double sided and must be proportioned so that there is no possibility of local buckling within the stiffener, i.e. stiffener outstands should be proportioned so that the local buckling reduction factor is equal to one.

Stiffener design is largely based on empirical relationships which tend to be somewhat conservative. However, the major cost in providing a stiffener arises from the fabrication costs of attaching the stiffener, rather than from the material costs of the stiffener itself, so that there is little to be gained by careful optimization of the stiffener dimensions. The basic requirement is simply that the stiffener should remain straight until collapse of the girder has occurred.

6.1 Transverse stiffeners

Intermediate transverse stiffeners play an important role in allowing the full ultimate load capacity of a girder to be achieved. In the first

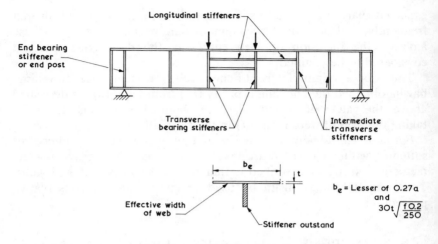

Fig. 8. Details of web stiffeners.

place, they increase the buckling resistance of the web; secondly, they must continue to remain effective after the web buckles to support the tension field; finally, they must prevent any tendency for the flanges to move towards one another.

These stiffeners must, therefore, be designed to satisfy both a stiffness and a stability check. In making either check, an effective section is taken consisting of the stiffener together with an effective width (b_e) of web plate, see Fig. 8, specified as:

$$b_e = \text{lesser of } 0 \cdot 27a \quad \text{and} \quad 30t\sqrt{\frac{f_{0 \cdot 2}}{250}} \tag{20}$$

To satisfy the stiffness check, the required minimum moment of inertia of the effective stiffener section is determined from linear buckling theory and empirical observation, as follows:

$$I_S \geq dt^3(2d/a - 0 \cdot 7) \tag{21}$$

where I_S is the moment of inertia of the effective stiffener section about an axis through its centroid parallel to the web.

To satisfy the stability check, the stiffener strut is designed as a typical compression member according to Clause 5.3. It is required that the factored resistance of the stiffener strut is in excess of $\frac{1}{3}$ of the factored shear load.

6.2 Transverse bearing stiffeners

Transverse bearing stiffeners must be provided at all positions of externally applied loads or reactions. They are designed in the same way as the intermediate transverse stiffeners discussed above except that, in the strength check, the fully factored externally applied load, or reaction, must be taken into account. Thus, in this case it is required that the factored resistance of the stiffener strut is in excess of one-third of the factored shear load *plus* the factored externally applied action.

6.3 End bearing stiffeners

The photographs of complete girders in Fig. 2 clearly show the additional functions performed by the end bearing stiffener or 'end post'. In addition to the normal requirements for transverse bearing stiffeners described above, the end post has to withstand the lateral loading imposed upon it by the horizontal component of the tension field forces in the adjacent web panel, as in Fig. 9(a). Under this loading, the end post acts as a vertical beam, spanning between the top and bottom flanges.

Because the end post beam is bending in its weak direction, this may well prove to be a difficult design condition to satisfy. Where possible, the end of the girder should be allowed to project beyond the support, as in Fig. 9(b), so that the resulting end post beam, comprising of two flanges and a web, provides a greater resistance to the bending action.

Where this is not possible, because of restriction on space at the support, then consideration should be given to the solution shown in Fig. 9(c), where the end web panel is designed to its buckling capacity only, i.e. neglecting any tension field capacity. The pull of the tension field developed in the panel adjacent to the end panel is then again resisted by a substantial vertical beam. Of course, the disadvantage with this arrangement is that the shear capacity of the end panel, since its post-buckling capacity is neglected, will be considerably lower than that of adjacent panels and this in a region where the applied shear force may well be a maximum. To minimise this effect, the width of the end panel should be reduced, as in Fig. 9(c), to increase its buckling capacity.

None of the three solutions to the problem of end post design, shown in Fig. 9, is ideal. However, the designer can select that which is most appropriate to the particular situation that he has to deal with.

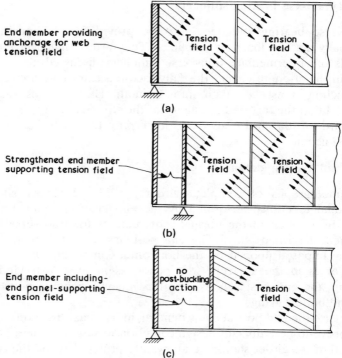

Fig. 9. Considerations for end post design.

6.4 Longitudinal stiffeners

As explained in Section 5, it is assumed that the main effect of the longitudinal stiffeners is to increase the initial buckling resistance of the web. After buckling, the tension field is assumed to develop over the complete web depth with the influence of the longitudinal stiffeners upon post-buckling behaviour being neglected.

Consequently, a stability check of the longitudinal stiffener as an idealised strut, as in the case of the transverse stiffeners, is not required. The longitudinal stiffener must simply meet the following empirical stiffness requirement (which is similar to that for transverse stiffeners in eqn (21)):

$$I_S \geq at^3(2a/d'_{av} - 0{\cdot}7) \tag{22}$$

where d'_{av} is the mean of individual sub-panel depths on either side of the stiffener.

I_S is the moment of inertia of the effective stiffener section about an axis through its centroid parallel to the web. As in the case of transverse stiffeners, the effective stiffener section is considered to contain an effective width b_e of web where:

$$b_e = \text{lesser of } 0.27 d'_{av} \quad \text{and} \quad 30 t \sqrt{\frac{f_{0.2}}{250}} \qquad (23)$$

7 LARGE OPENINGS IN WEBS (CLAUSE 5.6.6)

The provision of service ducts in buildings frequently requires openings to be provided in web panels. Although there is some experimental evidence of post-buckling action in steel plate girders with web openings, this was not considered adequate for the formulation of design clauses; the new code for the design of steel buildings (BS 5950) therefore limits the capacity of such webs to the initial buckling capacity. There is no available experimental evidence of the behaviour of aluminium webs with openings, so that the capacity is again limited to that at which buckling will first occur, neglecting any tension field capability.

The introduction of an opening will, of course, reduce the buckling capacity of the web plate and this reduction is represented by the introduction of a factor e_r where:

for a centrally positioned circular opening of diameter D,

$$e_r = \text{lesser of } 1 - \frac{D}{d} \quad \text{or} \quad \frac{D}{d} \qquad (24)$$

and for a centrally positioned rectangular opening of width b_o and depth d_o,

$$e_r = \text{lesser of } 1.24 - 1.16\left(\frac{d_o}{d}\right) - 0.17\left(\frac{b_o}{d_o}\right) \quad \text{or} \quad \frac{d_o}{d} \qquad (25)$$

Then, for an unstiffened web, the factored resistance is obtained from eqn (1) as:

$$V_{RS} = e_r \frac{V_1}{\gamma_m} C_S \qquad (26)$$

with C_S being determined from eqn (6).

Similarly, for a stiffened web, eqn (1) gives the factored resistance as:

$$V_{RS} = e_r \frac{V_1}{\gamma_m} C_{ST} \qquad (27)$$

and since the post-buckling capacity is now being neglected eqn (11) gives

$$C_{ST} = v_1 \qquad (28)$$

8 GIRDERS UNDER COMBINED SHEAR AND BENDING (CLAUSE 5.6.7)

In all the clauses discussed in this chapter so far, the capacity of a plate girder in resisting pure shear has been considered. However, girder panels are normally subjected to bending moments in addition to shear forces and the effects of a co-existent bending moment upon shear capacity are rather complex. The presence of a bending moment requires three additional factors to be considered in the determination of shear capacity:

(a) the reduction in the shear buckling stress of the web due to the presence of bending stresses;
(b) the influence of the bending stresses upon the magnitude of the tension field stresses required to produce yield in the web;
(c) the reduction of the plastic moment capacity of the flanges as a result of the axial flange stresses arising from the bending moment.

Although suitable for a computer solution, the equations resulting from these additional considerations do not reduce readily to a convenient design formula. Therefore, in the Code, the capacity of a girder to sustain the combined affects of bending and shear is represented by the simplified interaction diagram shown in Fig. 10. In this diagram, the shear capacity is plotted on the vertical axis and the moment capacity is plotted horizontally. The curve thus represents a failure envelope with any point lying upon it defining the co-existent values of factored bending and shear loads that will cause failure.

Point S on the vertical axis represents the pure shear capacity of the girder and is, therefore, determined from Clauses 5.6.2–5.6.4, as discussed earlier in this chapter. Point D on the horizontal axis

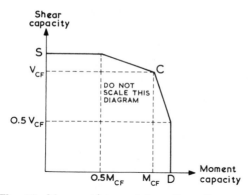

Fig. 10. Moment/shear interaction diagram.

represents the pure moment capacity of the girder, determined as for a beam in Clauses 5.5.2–5.5.5; these clauses for beams have been discussed in an earlier chapter.

At the intermediate point C, the mode of failure changes from a shear web mode to a flange bending mode. The point is determined on the familiar premise of assuming the web to carry all the shear and the flanges to carry all the bending moment.

Thus, the vertical ordinate of point C(V_{CF}) is determined as the web shear capacity, neglecting any contribution from the flange, i.e. from eqn (11),

$$V_{CF} = \frac{(v_1 + v_2)V_1}{\gamma_m} \qquad (29)$$

For transversely stiffened girders, values of v_1 and v_2 can be determined directly from the curves given in Figs 5(a,b).

The horizontal co-ordinate of point C (M_{CF}) is determined by assuming that the plate girder acts as a simple beam but ignoring the contribution of the web to the moment of resistance. M_{CF} thus represents the moment capacity of the flanges only.

The interaction diagram is completed by assuming that a moment of up to $0.5 M_{CF}$ will have no influence upon the pure shear capacity and that a shear of up to $0.5 V_{CF}$ will have no influence upon the pure bending capacity.

In this way, a full interaction diagram can be constructed very simply to allow the designer to determine the combined moment and shear capacity of his girder.

9 COMPARISON OF BS 8118 AND CP 118 VALUES

A detailed comparison of BS 8118 and the existing code CP 118 for aluminium girders has been conducted by the author. This parametric study of some 1368 girders has been fully reported elsewhere[11] but some of the more important results and conclusions will be summarised here.

Figure 11 shows the results for unwelded transversely stiffened girders of 5083 alloy with the ratio of the shear capacity determined from BS 8118 to the permissible shear calculated from CP 118 being plotted in the form of a histogram. For the maximum flange strength permitted in BS 8118, the predicted shear capacity is observed to be generally from 1·3 to 1·4 times that of CP 118. For a flange of zero strength, which approximates to the minimum flange rigidity allowed by CP 118, the ratio varies from 1·4 down to 0·8, thus showing CP 118 to have been rather unconservative in the case of weak flanges.

A similar comparison is presented in Fig. 12 for unwelded transversely stiffened girders of 6082 alloy. In this case the greater material

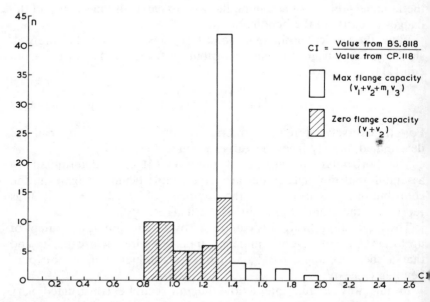

Fig. 11. Comparison of shear capacities from BS 8118 and CP 118—alloy 5083—unwelded.

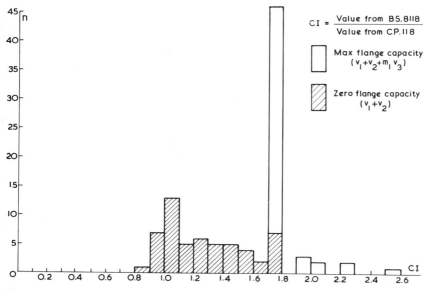

Fig. 12. Comparison of shear capacities from BS 8118 and CP 118—alloy 6082—unwelded.

strength allows greater benefits to be derived from BS 8118 when the maximum permissible flange strength is employed since the ratio is observed to be around 1·7–1·8 in such cases.

However, Fig. 13 shows the severe penalty that has to be paid because of the effects of welding on the 6082 alloy. These were not explicitly allowed for in the determination of permissible shear in CP 118 but the substitution of a value of the HAZ softening factor $w = 0.5$ into eqn (2) for V_1 and eqn (18) for C_{ST} has a very significant effect on the BS 8118 values.

It should, of course, be appreciated that the effects of welding are by far more significant in the 6082 alloy than in any other, so that the results summarised in Fig. 13 are not entirely representative of the comparison between BS 8118 and CP 118 values. Indeed, the general conclusion drawn from the detailed parametric study[11] was that the capacities derived from BS 8118 were almost invariably higher than those of CP 118. The main difference between the two codes is that BS 8118 allows the designer to take much greater advantage of the reserves of strength arising from the post-buckling action of a plate girder.

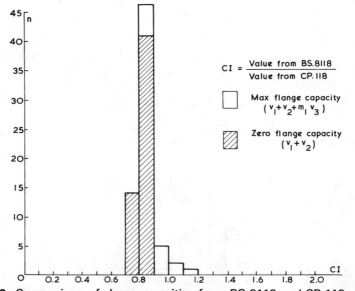

Fig. 13. Comparison of shear capacities from BS 8118 and CP 118—alloy 6082—welded.

10 COMPARISON OF BS 8118 AND EXPERIMENTAL VALUES

Little experimental information has been available until recently on the collapse behaviour of aluminium plate girders. A detailed study reported by Evans & Hamoodi[4] and a further study reported by Seah[5] has now provided some experimental evidence of the behaviour of both transversely stiffened and longitudinally stiffened girders under a variety of loading conditions. All these tests were carried out on welded girders of 6082 alloy, so that the HAZ softening effects were severe, and reference has already been made in this chapter to the mode of failure observed, see Fig. 4(b). It is the object of this present section simply to summarise the results obtained without repeating the detailed discussion of Refs 4 and 5.

Summarized results for the transversely stiffened girders are now presented in Table 1 and corresponding values for longitudinally stiffened girders are given in Table 2. All these girders were subjected to a predominant shear loading. The important girder parameters of

Table 1. Comparison of Code Values with Experimental Results for Transversely Stiffened Girders

Girder designation	Girder parameters			Terms in eqn (18)					$\frac{V_1(kN)}{eqn\ (2)}$	$\frac{V_{RS}(kN)^a}{eqn\ (1)}$	V_{EXP} (kN)	$\frac{V_{EXP}}{V_{RS}}$
	a/d	$\frac{d}{t}\sqrt{\frac{f_{0.2}}{250}}$	$\frac{v_1}{w}$	v_2	v_3	m^*	C_{ST}					
AG4	1·47	282	0·08	0·22	1·28	0·112	0·443		61·4	27·3	52·3	1·92
AG1	1·47	284	0·06	0·22	1·28	0·172	0·500		59·5	29·8	56·3	1·89
AG5	1·46	284	0·06	0·22	1·28	0·220	0·562		60·9	34·1	69·3	2·03
AG3	1·47	376	0·02	0·23	1·28	0·136	0·424		41·1	17·3	33·9	1·96
AG2	1·47	376	0·02	0·23	1·28	0·207	0·505		41·4	20·9	41·4	1·98
AGS1	0·97	399	0·08	0·31	1·70	0·119	0·587		43·9	25·8	47·0	1·82
AGS2-T1	0·48	382	0·24	0·43	2·18	0·125	0·935		39·8	37·2	62·0	1·67
AGS2-T2	0·48	383	0·24	0·43	2·18	0·124	0·935		39·9	37·3	63·5	1·70
AGS3-T2	0·48	382	0·24	0·43	2·18	0·124	0·935		39·8	37·2	59·0	1·59
AGS4-T1	0·48	382	0·24	0·43	2·18	0·126	0·935		39·8	37·2	60·5	1·63
AGCS1	0·48	392	0·24	0·43	2·18	0·130	0·955		42·2	40·3	58·8	1·46
AGCS2	0·48	392	0·24	0·43	2·18	0·130	0·955		42·1	40·2	61·8	1·54
AGCS3	0·48	392	0·24	0·43	2·18	0·130	0·955		42·1	40·2	58·7	1·46

Information on Girders AG4 to AGS4-T1 obtained from Ref. 4.
Information on Girders AGCS1 to AGCS3 obtained from Ref. 5.
[a] Note that V_R has been calculated from eqn (1) assuming $\gamma_m = 1$ since the material properties were accurately determined in the experiments.

Table 2. Comparison of Code Values with Experimental Results for Longitudinally Stiffened Girders

Girder designation	Girder parameters		Terms in eqn (18)					$V_1(kN)$ eqn (2)	$V_{RS}(kN)$[a] eqn (1)	V_{EXP} (kN)	$\dfrac{V_{EXP}}{V_{RS}}$
	a/d	$\dfrac{d'}{t}\sqrt{\dfrac{f_{0.2}}{250}}$	$\dfrac{v_1}{w}$	v_2	v_3	m^*	C_{ST}				
AGLS2	1·47	133	0·30	0·22	1·30	0·125	0·680	39·7	27·1	44·0	1·62
AGLS3	1·47	133	0·30	0·22	1·30	0·124	0·680	39·7	27·0	43·5	1·61
AGLS4	1·47	133	0·30	0·22	1·30	0·126	0·681	44·2	30·1	46·5	1·55
AGLS5	1·47	100	0·60	0·19	1·19	0·125	0·938	38·5	36·1	49·5	1·37
AGCS1	1·49	129	0·36	0·20	1·24	0·130	0·721	41·6	30·0	41·8	1·40
AGCS2	1·51	127	0·36	0·20	1·23	0·130	0·720	41·1	29·6	45·0	1·52
AGCS3	1·52	127	0·36	0·20	1·23	0·130	0·721	41·2	29·7	45·4	1·53

Information on Girders AGLS2 to AGLS5 obtained from Ref. 4.
Information on Girders AGCS1 to AGCS3 obtained from Ref. 5.
[a] Note that V_R has been calculated from eqn (1) assuming $\gamma_m = 1$ since the material properties were accurately determined in the experiments.

aspect ratio (a/d) and web slenderness $\left(\dfrac{d}{t}\sqrt{\dfrac{f_{0\cdot2}}{250}}\right)$ are listed in each case, together with the calculated values. The value of each term in eqn (18) is tabulated to give an indication of the relative importance of the different sources from which each girder derives its strength.

The value of the shear resistance (V_{RS}) calculated according to eqn (1) has been determined assuming the partial safety factor on material strength γ_m to be equal to 1·0, since the material properties of the test girders were accurately determined. Typically, γ_m should be taken as 1·25 for such welded girders. The value of the experimentally determined failure load (V_{EXP}) is tabulated for each girder for comparison with the code value and the ratio of V_{EXP}/V_R is used to represent the degree of comparison achieved.

For all the 20 girders considered, the value of the ratio is seen to be considerably higher than 1·0. Indeed, in some cases, the experimentally measured capacity lies between 1·5 times and twice the capacity determined from the code. Whilst confirming the most important fact that the code predictions are, therefore, safe for design, the high values of the V_{EXP}/V_{RS} ratio do indicate considerable conservatism in the code values. This conservation arises from the treatment of HAZ softening, and it should be appreciated that all these experimental girders were made of 6082 alloy where softening effects are particularly severe ($w = 0\cdot5$). Given that many of the observed failures were of a rather 'brittle' nature, involving fracture of the web plate, see Fig. 4(b), the degree of conservatism in the code prediction is not considered to be excessive until further research information has become available.

11 CONCLUSION

This chapter has described in some detail the procedures specified in the new code for determining the resistance of plate girders. Emphasis has been placed upon the general procedures and philosophy as well as on the detailed equations given in the code; the underlying theory has also been outlined.

The code allows the designer to take advantage of reserves of strength arising from post-buckling behaviour whilst retaining a physical feel for the structural action and without requiring complex calculations which could discourage innovation and optimisation of design. Such a compromise has been achieved by simplifying the analytical approach and this has inevitably introduced some degree of conservatism, as indicated by the comparisons with recently obtained experimental data.

ACKNOWLEDGEMENTS

Some of the work described in this chapter was supported by the Royal Armament Research and Development Establishment and the author wishes to acknowledge this support. He is also grateful for many helpful discussions with Dr P. S. Bulson, Mr J. B. Dwight, Professor G. A. O. Davies, Dr. O. Vilnay and Miss C. A. Burt on various aspects of the code.

REFERENCES

1. British Standards Institution, *Code of Practice for Design of Steel Bridges, BS 5400 : Part 3*. BSI, London, 1982.
2. British Standards Institution, *The Structural Use of Steelwork in Buildings*, BS 5950. BSI, London, 1985.
3. Evans, H. R., Longitudinally and transversely reinforced plate girders. In *Plated Structures, Stability and Strength,* Ch. 1. Applied Science Publishers, London, 1983.
4. Evans, H. R. & Hamoodi, M. J., The collapse of welded aluminium girders—an experimental study. *Thin-walled Structures,* **5**(4) (1987) 247–275.
5. Seah, H. A., The behaviour of welded aluminium alloy plate girders reinforced with carbon fibre reinforced plastics. MSc thesis, University College, Cardiff, 1984.
6. British Standards Institution, *The Structural Use of Aluminium*, CP 118. BSI, London, 1969.
7. Dubas, P. & Gehri, E. (eds), Behaviour and design of steel plated structures. ECCS—Technical Working Group 8.3. Publication No. 44, Zurich 1986.
8. Hoglund, T., Simply supported long thin plate I-girders without web stiffeners, subjected to distributed transverse load. *Proceedings, IABSE Colloquium,* London, 1971.
9. Allen, H. G. & Bulson, P. S., *Background to Buckling.* McGraw-Hill, Maidenhead, 1980.
10. Evans, H. R., Porter, D. M. & Rockey, K. C., A parametric study of the collapse behaviour of plate girders. University College, Cardiff, Research Report, 1976.
11. Evans, H. R., Comparison of draft code BS 8118 with CP 118 for the design of aluminium plate girders. University College, Cardiff, Research Report, CAR/RARDE/1, 1985.

5

Welded Connections

F. SOETENS

TNO Building and Construction Research, Delft, The Netherlands

ABSTRACT

The results of a research programme on welded connections in statically loaded aluminium alloy structures are reported. The motives for research on the structural behaviour of aluminium alloys are given and the research programme is briefly described, then results of the various parts of the research programme are discussed and evaluated.

1 INTRODUCTION

Although much research on aluminium alloys has been carried out in the past, relatively little attention has been given to their structural behaviour. Therefore, most design rules for aluminium alloy structures are based on design rules for steel structures, as illustrated, for example, by the ECCS Recommendations,[1] DIN 4113[2] and TGB-Aluminium.[3]

Applying similar design rules is permissible, since aluminium and steel structures do have very similar structural behaviour. This means, in particular, the application of limit state design methods which allow a better description of the real (non-linear) behaviour of a structure than allowable stress methods. Knowledge is required of the structural behaviour of the members, and apart from strength and stiffness this behaviour is determined by the deformation capacity which enables redistribution of forces to occur in a structure.

The aim of this study was to investigate the structural behaviour of aluminium alloy members and in particular the deformation capacity, in order to realize full application of limit state design methods.

It was decided to investigate welded connections in statically loaded aluminium alloy structures for some of the reasons listed below:

- To realize full application of limit state design methods.
- Welded connections are well suited for studying the relationship between mechanical properties and structural behaviour, since stress and strain concentrations occur.
- Design rules based on ultimate limit states are concerned with failure of the structure. In spite of many investigations carried out, characteristic strength values of weld metal and heat-affected zone, respectively, do not exist.
- There are no uniform design rules for butt welds, fillet welds and welded connections in aluminium alloys. The use of rules similar to those for steel has to be investigated.
- Connections form an important part of the cost of structures, so simple, unstiffened connections are preferred. However, this means that stress and strain concentrations are introduced which have to be studied in relation to the strength, stiffness and deformation capacity of the joint.

It is of interest to investigate welded connections in statically loaded aluminium alloy structures because design rules for aluminium are more restrictive than those for steel and also because many design codes differ considerably from one another in the rules they give.

2 RESEARCH PROGRAMME FOR WELDED CONNECTIONS

In September 1981 the research programme 'Welded connections in aluminium alloy structures' was started, and it was completed in December 1984. It was carried out by TNO together with Alcoa Nederland, Alurage, Bayards, ICB and UCN, and supported by the Netherlands Government.

The programme was subdivided as follows:

State of the art

In this literature study the materials, i.e. the alloys and filler metals, the welding technology, and the mechanical properties of welds and welded connections, were reviewed.

Theoretical and experimental research
This part of the programme was subdivided into:

—*Experimental research on mechanical properties.* In this study the mechanical properties of weld metal and heat-affected zone were investigated.
—*Experimental research on fillet welds.* In this study the strength of fillet welds was investigated and the design of fillet welds in aluminium alloy structures was considered.
—*Theoretical and experimental research on welded connections in hollow sections.* In this study X- and T-connections of rectangular hollow sections were investigated, with particular reference to the relation between the mechanical properties and structural behaviour of the joint.

Evaluation
Finally, the results of the programme were evaluated and recommendations given for the design of welded connections in statically loaded aluminium alloy structures.

3 STATE OF THE ART

The results of the literature study are reported in Ref. 4. The most important results are summarized below.

3.1 Aluminium alloys and welding technology

Aluminium alloys of the series 5xxx, 6xxx and 7xxx are commonly used for load-bearing structures. The 5xxx series is mainly associated with rolled material, while the 6xxx and 7xxx series are more likely to be concerned with extrusions. Material thicknesses are usually smaller than $t = 20$ mm, many applications can be found in thin-walled structures with thicknesses below $t = 6$ mm.

The filler metals most frequently used are: 5356 (=Al Mg 5), 5183 (=AlMg 4·5 Mn) and 4043 (= AlSi 5). Filler metal 5356 can be combined with nearly all parent metal alloys and is therefore the most commonly applied. For some combinations filler metal 5183 is preferred to 5356, while filler metal 4043 is mostly combined with parent metal of the 6xxx series.

As far as the welding processes are concerned, the MIG (= Metal Inert Gas welding) and the TIG (= Tungsten Inert Gas welding) processes are generally applied. With the MIG process the following can be distinguished: spray-arc, pulsed-arc and sometimes plasma—MIG. Short-circuit MIG is not used because of poor results for the welding of aluminium alloys. With TIG welding, mainly TIG-AC (= Alternating Current) is used.

3.2 Mechanical properties of weld metal and heat-affected zone

Mechanical properties of weld metal are often given for prescribed combinations of parent and filler metal. For the heat-affected zone of non-heat-treatable alloys in the strain-hardened condition the mechanical properties of the parent metal in the annealed condition are applied. For heat-treatable alloys mechanical properties in the heat-affected zone are given for a limited number of treatments (often only T6).

In Ref. 2 allowable stresses are given both for the weld metal and the heat-affected zone. In many standards only the lowest allowable stress or the design strength of the weld metal and heat-affected zone is given (see Refs 1, 3, 5 and 6). In addition, as shown in Table 1, the values for the mechanical properties differ significantly according to Refs 1, 2 and 5.

It can be inferred from the results of investigations carried out in recent years (see Refs 7–10), that as far as the weld metal is concerned, the mechanical properties depend on the combination of parent metal and filler metal, the welding process, the plate thickness

Table 1. Allowable Stresses According to ECCS,[1] DIN 4113[2] and CP 118[5]

Allowable stresses weld metal/heat-affected zone (N/mm^2)

Parent metal	7020-T6		6082-T6		6060-T5	5083-0	5454/5754 H24
Filler metal	5356	4043	5356	4043	5356	5356	5356
ECCS	112	82	75	75	50	80	50
DIN 4113	95		75	75	45	75	45
CP 118	124		51	51	31	82	62

and the weld type. For the mechanical properties of the heat-affected zone the heat-input is of course very important. To give some idea of the variation in values, the main results for the ultimate strength (mean values!) of weld metal and heat-affected zone—according to Ref. 7—are shown in Table 2. It is noted that these results were derived from butt-welded test specimens with a thickness $t = 20$ mm.

3.3 Design of welds

3.3.1 Butt welds

The design of butt welds (see Fig. 1) presents few difficulties. The distribution of forces acting in the weld is similar to that in the connected members, and the stressed area is known (for fully-penetrated butt welds: the weld throat $a \geqslant$ thickness t). This means that the mechanical properties of the weld metal and heat-affected zone are the only design parameters.

3.3.2 Fillet welds

In addition to defining the mechanical properties of the weld metal and heat-affected zone, in the case of fillet welds the definition of the rupture section of the weld, and the definition of the stresses in that section, is difficult.

In Refs 1 and 2 for *transverse fillet welds*, in which the direction of the force is perpendicular to the axis of the weld (see Fig. 1), the rupture section (= throat section of the weld) is turned to the horizontal or the vertical position, parallel to the direction of the force. This gives a shear stress $\tau_\perp$ perpendicular to the axis of the weld. For *longitudinal fillet welds*, in which the direction of the force is parallel to the axis of the weld, the shear stress $\tau_\parallel$ (= force divided by the area of the throat section) governs the design.

Another approach which is frequently given in design rules is to divide the force or the resultants of all forces by the total weld throat area $\left(\dfrac{F}{\sum al}\right)$ independently of the orientation of the force and the weld. This means that, for reasons of simplicity, no distinction is made between the strength of transverse and longitudinal fillet welds. In Ref. 5 a similar approach is used, but the strengths of transverse and longitudinal fillet welds are distinguished. For transverse fillet welds allowable *tensile* stresses similar to those for butt-welded joints are given, while with longitudinal fillet welds the allowable *shear* stresses

Table 2. Main Results for Butt Welds According to IIW Research Programme[7]

Ultimate strength (tests[a]) heat-affected zone ($\sigma_{u_{HAZ}}$) and weld metal σ_{u_w} (N/mm²)

Parent metal	Filler metal	Germany MIG	Germany TIG	Italy TIG	N'lands TIG	Poland TIG	Remarks
6082-T6	4043	$\sigma_{u_w} = 227$	$\sigma_{u_w} = 164^b$ $\sigma_{u_{HAZ}} = 169$	$\sigma_{u_w} = 142^b$	$\sigma_{u_w} = 168$ $\sigma_{u_{HAZ}} = 175$	$\sigma_{u_w} = 178$	Mean values from tests
5083-F	5356	$\sigma_{u_w} = 309$ $\sigma_{u_{HAZ}} = 306$	$\sigma_{u_{HAZ}} = 304$	$\sigma_{u_w} = 274^b$	$\sigma_{u_{HAZ}} = 305$	$\sigma_{u_w} = 272$	See above
7020-T6	5356	$\sigma_{u_w} = 322$	$\sigma_{u_w} = 214^b$	$\sigma_{u_w} = 212^b$	$\sigma_{u_{HAZ}} = 220^c$	$\sigma_{u_w} = 203$	See above

[a] Tests on butt-welded connections, X-type weld and thickness $t = 20$ mm.
[b] Weld defects as lack of penetration, lack of fusion and porosity were observed.
[c] Failure heat-affected zone at a low stress level, caused by wrong pre-heating.

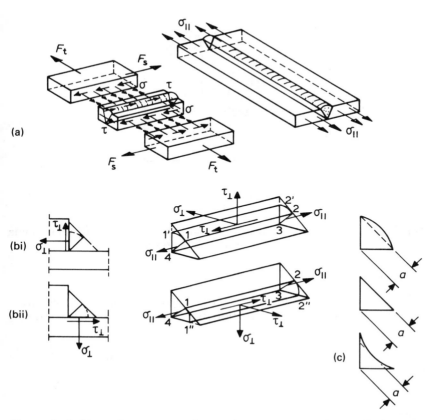

Fig. 1. Definition of rupture section and stresses according to Refs 1 and 2. Definition of welds and stresses in welds: (a) butt weld; (b) fillet weld, (bi) rupture section 1234 = throat section turned into the vertical position 1'2'34, (bii) rupture section turned into the horizontal position 1"2"34 (general case); (c) throat thickness for the calculation.

coincide. This means that the transverse fillet weld is presumed to be 60% stronger than the longitudinal fillet weld. In Ref. 1 a figure of 45% and in Ref. 2 a figure of 20% is given.

A third approach for determining the strength of fillet welds is to apply the β-formula which is commonly used for steel structures (IIW, ISO and ECCS Recommendations for steel). In Refs 3 and 6 the β-formula is prescribed, and this can be used successfully for aluminium alloy structures as shown in Ref. 11. With the β-formula it is

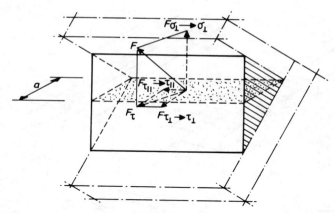

Fig. 2. Definition of stresses in minimum throat section of fillet welds according to Refs 3 and 6.

assumed that:

—failure occurs in the minimum throat section of the weld; the stress distribution in the throat section satisfies equilibrium; the stresses are defined in Fig. 2.

According to the β-formula the stresses in the throat section have to satisfy:

$$\beta\sqrt{\sigma_\perp^2 + 3(\tau_\perp^2 + \tau_\parallel^2)} \leq \sigma_d$$

where

σ_d = design strength of weld metal (in Refs 3 and 6: $\sigma_d = \sigma_{0,2_w}$),
β = factor to compensate for the difference in strength and ductility of fillet weld metal and butt weld metal, and to correct the value 3 which in reality is 2·6 for failure.

Applying the β-formula, a difference in design strength between transverse and longitudinal fillet welds of $\dfrac{\sqrt{3}}{\sqrt{2}} = 1\cdot22$ is assumed.

In Refs 7, 8 and 10 results from tests on transverse and longitudinal fillet welds are reported. It appears that:

- The rupture section is very close to the minimum throat section of the welds, except in the case of low strength parent metals and high-strength filler metal. In this case failure occurred at the legs of the weld, as was expected.

- Transverse fillet welds are at least 20% stronger than longitudinal fillet welds. An 'exact' figure cannot be derived from these tests.
- Except for parameters similar to those mentioned in Section 3.2 and influencing the mechanical properties of fillet welds, it was observed that the shear strength of longitudinal fillet welds was below the shear strength of the weld metal (butt weld test).
- The deviation in results was considerable, which meant a considerable difference between mean value and characteristic strength (95% confidence level).
- Many weld defects were found, such as lack of fusion, lack of penetration and porosity. MIG welding yielded better results than TIG, while longitudinal fillet welds showed fewer defects than transverse fillet welds.
- The results of Refs 7 and 8 are based on the rupture section, while for 10 the nominal section is considered. This makes a comparison of results difficult.

The results have been analysed and, where possible, characteristic values for the ultimate strengths have been determined. The main results from Ref. 7 and 8 are shown in Table 3, while comparable results from 8 and 10 are given in Table 4. However, it should be noted that these results have no general validity, since:

—The number of parameters varied is too small (in Ref. 7: two filler metals, two thicknesses $t = 12$ and 20 mm; in Ref. 8: one thickness $t = 8$ mm and only MIG welding; in Ref. 10: only MIG welding).
—The number of tests per parameter did not always allow a statistical evaluation of results; only mean values were given.

Because of this the conclusions and design values for the strength of both butt and fillet welds given in these studies cannot be endorsed.

3.4 Design of welded connections

For the design of welded connections it is often required that a prescribed allowable stress is nowhere exceeded in the connection. This approach is followed in Ref. 2, among others. Even with simple connections elaborate calculations have to be carried out, so simplifications are applied which lead to over-dimensioning of the connection. Besides, in reality the allowable stress will always be exceeded locally because of stress and strain concentrations.

Table 3. Ultimate Shear Strengths for Fillet Welds According to Refs 7 and 8, Characteristic and Mean Values (in Brackets) for Stresses $\tau_\parallel$ and $\tau_\perp$ in N/mm²

Welding processes Type of fillet weld Throat dimension	Parent metal	Results according to Ref. 7							Results according to Ref. 8 MIG	
		MIG					TIG				
		Longitudinal			Transv.	Longitudinal		Transv.	Long.	Transv.
		4 mm	7 mm	10 mm	7 mm	4 mm	7 mm	7 mm	4 mm	4 mm
Filler metal 5154	6082-T6								(52)	135(172)
	5754-H34								(127)	153(191)
5356	7020-T6	149	(121)	(155)	(203)	(140)	111(135)	106(142)	(164)	181(207)
	6082-T6								(142)	153(182)
	5754-H34								(150)	178(204)
	5083-F	161(214)	(157)	(168)	(160)	(152)	132(151)	89(155)		
5183	7020-T6								(167)	170(216)
	6082-T6								(72)	170(205)
	5754-H34								(144)	177(206)
4043	6082-T6	118(151)	(127)	(116)	(164)	(90)	72(92)	67(110)	(123)	125(151)

Table 4. Ultimate Shear Strengths for Fillet Welds According to Refs 10 and 8; Characteristic and Mean Values (in Brackets) for Stresses $\tau_\parallel$ and $\tau_\perp$ in N/mm² (Note: in columns 1 and 2 sometimes two values are given, the higher for $t = 4$ mm and the lower $t = 20$ mm!)

Welding process Type of fillet weld		Results according to Ref. 10 MIG		Results according to Ref. 8 MIG		Column 2 divided by column 1	Remarks
Parent metal	Filler metal	Long.	Transv.	Long.	Transv.		
		(1)	(2)	(3)	(4)	(5)	(6)
5754-H111	5154	81	107	(127)	153	1·32	Ref. 8: results for 5754-H34
5086-H111	5183	96	169/201	—	—	1·76	column 3 and 4 according to Ref. 7
	5356	103	148/228	161	(160)	1·44	
6060-T5	4043	71	132	—	—	1·86	
6060-T6	5356	95	135/167	—	—	1·42	
6082-T6	4043	82/90	100/152	118	125	1·22	
	5356	86	161	(142)	153	1·87	
7020-T6	4043	76/100	142	—	—	1·87	
	5356	110	187	149	181	1·70	
	5280	132/147	159/207	—	—	1·20	
7051-T6	4043	86/110	116	—	—	1·35	
	5356	102/126	116/148	—	—	1·14	
	5280	94/118	164/196	—	—	1·74	

A second approach is to distribute the forces acting on the connection over the respective welds in a convenient way and take care to satisfy equilibrium. Then a check should be made to see if the welds are capable of carrying the loads distributed to them. This approach is recommended in Refs 1 and 6.

A third approach is very similar to the second, but gives additional rules for the design, i.e. it should be checked whether the forces distributed over the welds can occur in reality. In other words, do the welds possess sufficient deformation capacity to allow the assumed force distribution over them?

This approach is given in Ref. 3. Alternatively, it is permissible to calculate the welds for the stresses occurring in the connected parts of a member; for example, in a beam-column connection loaded by moment and shear forces, the flange welds balance the tensile stresses in the flange due to the moment force, and the web welds balance the tensile and shear stresses due to both moment and shear forces, see Ref. 11.

The results of this literature study have determined the parameters to be studied in the experimental and theoretical research described in the following sections.

4 EXPERIMENTAL RESEARCH ON MECHANICAL PROPERTIES

4.1 Choice of parameters

In view of the results of the literature study concerning mechanical properties of the weld metal and heat-affected zone, it was decided to carry out an experimental research study into the parameters which really influence the mechanical properties. The following parameters were chosen:

- four alloys, namely, 5083-0, 6063-T5, 6082-T6 and 7020-T6;
- two filler metals, namely, 5356 and 4043;
- two plate thicknesses, namely, $t = 4$ mm and $t = 12$ mm;
- two weld types, namely, T-weld (square-groove weld) and V-weld (single V-weld);
- two welding processes, namely, MIG (spray-arc) and TIG (AC).

In this programme, in which all specimens were welded by one firm, the following were not varied: the welder, the welding position (1G) and the welding parameters (determined by the weld quality which was chosen at level 2AaAb according to the IIW classification of radiographs, see also Section 4.2).

In order to restrict the number of tests and because all combinations of parameters did not need to be investigated (for instance: $t = 4$ mm and a V-weld), the combinations shown in Tables 5 and 6 were studied.

Note: 1. The plate thicknesses and the weld types are combined.
2. Filler metal 4043 is not combined with alloy 5083.
3. TIG welding is only used for thin plates.

In addition to the above test specimens, which were welded by Bayards, butt-welded plates were prepared by other firms participating in this research programme (see Section 2). In the additional programme the following were investigated: alloys 6063 and 6082 (and additionally: 1060, 5005 and 5052); filler metals 5356 and 4043 (and additionally: 5554); plate thickness $t = 4$ mm; TIG and MIG welding.

4.2 Welding test specimens

The test specimens were welded according to a detailed specification concerning edge preparation, cleaning, tacking, qualification of welding procedure and welder, and recommendations for the welding parameters to be used. Also approval of the welding procedure, welder and choice of welding parameters based on the results of radiographic testing which were compared with the reference radiographs of IIW.

The specimens used in this programme were butt-welded plates as shown in Fig. 3. In this programme qualification specimens as well as test specimens were used, provided both received a similar classification. If this classification deviated from 2AaAb, the location and the dimension of the origin of this deviation (the 'weld defect') was recorded on the plates. A period of 6 weeks separated welding and testing.

Table 5. Ultimate strengths σ_u (N/mm^2) of Tensile Specimens $\perp$ to the Weld Axis (see Fig. 3). Failure of Heat-Affected zone: $\sigma_{u_{HAZ}}$. Failure of Weld: σ_{u_w}

Welding process	Filler metal	Thickness	5083-0	6063-T5	6082-T6	7020-T6
MIG	5356	$t = 4$ mm	$\sigma_{u_{HAZ}} = 296$ 295 298 298 298	$\sigma_{u_{HAZ}} = 150$ 150 151 152 149	$\sigma_{u_{HAZ}} = 206$ 216 210 212 216	$\sigma_{u_w} = 265$ 244 251 260 253
		$t = 12$ mm	$\sigma_{u_{HAZ}} = 304$ 318 313 323 320	$\sigma_{u_{HAZ}} = 159$ 157 159 162 158	$\sigma_{u_{HAZ}} = 223$ 222 224 228 216	$\sigma_{u_w} = 299$ 284 296 305 302
	4043	$t = 4$ mm		$\sigma_{u_{HAZ}} = 151$ 151 154 152 151	$\sigma_{u_{HAZ}} = 216$ 198 217 210 214	$\sigma_{u_w} = 224$ 245 248 239 244
		$t = 12$ mm		$\sigma_{u_{HAZ}} = 155$ 150 151 151 152	$\sigma_{u_{HAZ}} = 217$ 225 226 222 224	$\sigma_{u_w} = 217$ 243 223 229 239
TIG	5356	$t = 4$ mm		$\sigma_{u_{HAZ}} = 142$ 154 141 140	$\sigma_{u_{HAZ}} = 167$ 169 174 171	

Table 6. Ultimate strengths σ_u (N/mm^2) of Tensile Specimens || to the Weld Axis. Failure of Heat-Affected Zone: σ_{uHAZ}, Failure of Weld: σ_{uw}

Welding process	Filler metal	Thickness	5083-0	6063-T5	6082-T6	7020-T6
MIG	5356	t = 4 mm	σ_{uw} = 245 258 255 259 251	σ_{uw} = 177 174 176 173 174	σ_{uw} = 232 223 238 229 231	σ_{uHAZ} = 390 384 385 379 382
MIG	5356	t = 12 mm	σ_{uw} = 277 278 285 258 268	σ_{uw} = 217 221 217 218 212	σ_{uw} = 230 240 236 241 228	σ_{uHAZ} = 409 422 420 420 400
MIG	4043	t = 4 mm		σ_{uw} = 168 172 172 150 169	σ_{uw} = 210 219 208 205 212	σ_{uHAZ} = 384 381 379 367 378
MIG	4043	t = 12 mm		σ_{uw} = 178 182 187 186 180	σ_{uw} = 222 210 211 205 212	σ_{uHAZ} = 411 417 416 414 415
TIG	5356	t = 4 nm		σ_{uw} = 149 186 184 192 185	σ_{uw} = 227 218 209 230 236	

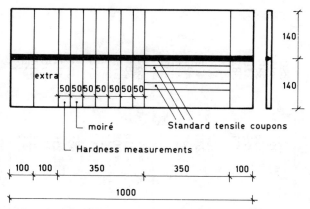

Fig. 3. Butt welded joint; dimensions in mm; position of the weld; position and dimensions of different test specimens.

4.3 Test results

To determine the mechanical properties of the parent metal, weld metal and heat-affected zone, standard tensile coupons taken parallel to the weld axis, as shown in Fig. 3, were tested. The results have been summarized in Table 7.

A macro section of all the welded joints was made perpendicular to the weld axis to measure the hardness of the parent metal, heat-affected zone and weld metal according to the Brinell Method. One example is shown in Fig. 4. All the results of the hardness measurements, including those of the fillet welded joints (see Section 5), are given in Ref. 12.

To determine the ultimate strength of the weld metal and heat-affected zone with sufficient reliability and to check the influences on these strengths mentioned in the literature, at least five tensile specimens were taken from each butt-welded joint perpendicular to the weld axis, as shown in Fig. 3. Specimens with weld defects (see Section 4.2) were included in these five specimens. The weld reinforcement of the tensile specimens was removed.

The results of these tensile tests are given in Table 5. For the alloys 5083, 6063 and 6082 failure occurred in the heat-affected zone, while for alloy 7020 weld failure occurred.

To determine the ultimate strength of the weld metal for the alloys 5083, 6063 and 6082 and the ultimate strength of the heat-affected

Table 7. Typical Mechanical Properties of the Parent Metal, Heat-Affected Zone and Weld Metal

(a) Parent metal

Thickness	Mech. prop.	5083-0	6063-T5	6082-T6	7020-T6
t = 4 mm	$\sigma_{0,2}$ (N/mm^2)	142	178	284	408
	σ_u (N/mm^2)	300	218	316	442
	A5 (%)	26	12	13	14
t = 12 mm	$\sigma_{0,2}$ (N/mm^2)	184	188	310	404
	σ_u (N/mm^2)	334	210	338	448
	A5 (%)	19	18	14	12

(b) Heat-affected zone

Welding process	Thickness	Mech. prop.	5083-0	6063-T5	6082-T6	7020-T6
MIG	t = 4 mm	$\sigma_{0.2}$ (N/mm^2)	150	134	166	276
		σ_u (N/mm^2)	298	172	238	388
		A5 (%)	25	14	14	12
	t = 12 mm	$\sigma_{0.2}$ (N/mm^2)	160	136	208	262
		σ_u (N/mm^2)	320	172	256	400
		A5 (%)	21	18	13	14
TIG	t = 4 mm	$\sigma_{0.2}$ (N/mm^2)		88	118	
		σ_u (N/mm^2)		150	190	
		A5 (%)		17	16	

(c) Weld metal

Welding process	Filler metal	Thickness	Mech. prop.	5083-0	6063-T5	6082-T6	7020-T6
MIG	5356	t = 4 mm	$\sigma_{0.2}$ (N/mm^2)	134	87	128	174
			σ_u (N/mm^2)	240	174	220	258
			A5 (%)	13	14	13	7
		t = 12 mm	$\sigma_{0,2}$ (N/mm^2)	128	99	106	170
			σ_u (N/mm^2)	268	212	228	296
			A5 (%)	17	22	16	6
	4043	t = 4 mm	$\sigma_{0,2}$ (N/mm^2)		89	116	170
			σ_u (N/mm^2)		156	204	239
			A5 (%)		9	9	4
		t = 12 mm	$\sigma_{0,2}$ (N/mm^2)		83	108	140
			σ_u (N/mm^2)		180	204	229
			A5 (%)		13	9	4
TIG	5356	t = 4 mm	$\sigma_{0,2}$ (N/mm^2)		104	101	
			σ_u (N/mm^2)		204	227	
			A5 (%)		16	14	

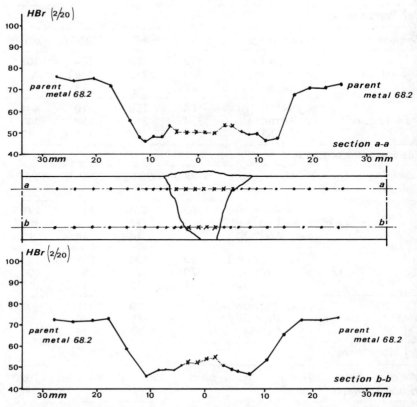

Fig. 4. Brinell hardness measurements. Butt-welded specimen, thickness $t = 12$ mm. Alloy: 6063-T5. Filler metal: 4043. Welding process: MIG.

zone for the alloy 7020, five tensile coupon specimens were taken from each butt-welded joint parallel to the weld axis. The results of these tests are given in Table 6.

Besides the ultimate strengths, the deformations occurring both in the heat-affected zone and weld were measured using the moiré method. One tensile specimen perpendicular to the weld axis was taken from each welded joint, see Fig. 2. Moiré photographs of the specimen [6063-T5/5356/4T/TIG-welded] are shown in Fig. 5. Failure occurred in the heat-affected zone. Specimen [7020-T6/5356/12V/MIG-welded] is shown in Fig. 6, and failure is seen to have occurred in the weld.

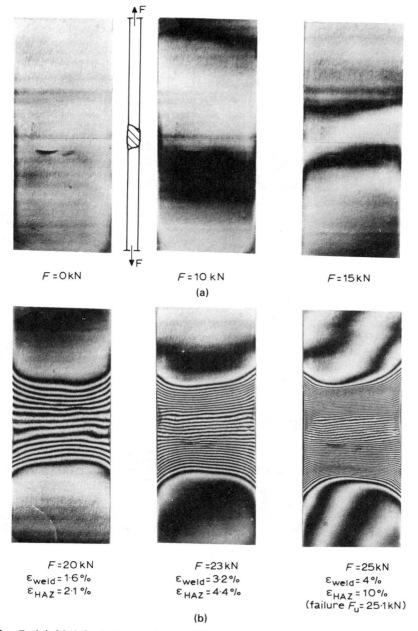

Fig. 5. (a) Moiré photographs at different load levels, specimen 6063-T5/5356/4T/TIG; scale 1:1, undeformed grid: 20 lines/mm, so each moiré line corresponds to a displacement of 0·05 mm. (b) Moiré photographs at load levels close to failure, failure of the heat-affected zone.

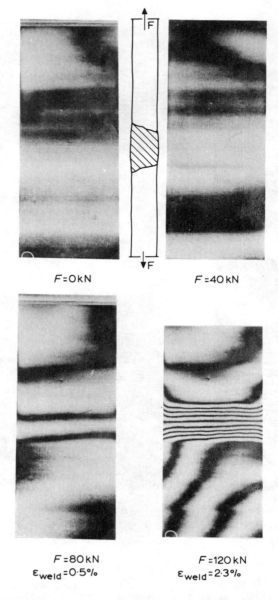

Fig. 6. (a) Moiré photographs of specimen 7020-T6/5356/12V/MIG; scal 1:1.

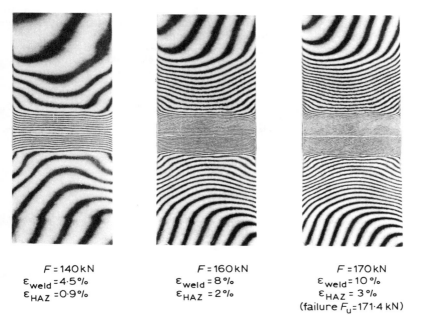

Fig. 6—*contd.* (b) Moiré photographs at load levels close to failure, failure of the weld.

The results of tensile tests for determining the ultimate strength of the weld metal and heat-affected zone, and for measuring the deformations, are given in Ref. 13.

4.4 Evaluation of results

The results of the ultimate strengths of the weld metal and heat-affected zone (Tables 5 and 6) have been evaluated statistically by means of an analysis of variance (see Ref. 14). The influence of parameters varied (or combinations of parameters) on the total deviation of the test results is measured by this method. In Table 8 the analysis of variance as determined for the ultimate strength of the weld metal is given, and similarly in Table 9 for the ultimate strength of the heat-affected zone. By comparing the variances due to the different sources and listed in the last column of Tables 8 and 9 with the residual variance, which is an estimate of the magnitude of the

Table 8. Table of Analysis of Variance of the Ultimate Strengths of the Weld Metal σ_{u_w} (Tables 5 and 6). Parameters Varied Were: Alloy, A; Filler Metal, F; Thickness, t.

Nature of effect	Source	Sum of squares	Degrees of freedom	Variance estimate
Main parameters	A	49 534	2	24 767
	F	11 677	1	11 677
	t	3 921	1	3 921
Interaction between two parameters	AF	1 247	2	624
	At	1 755	2	878
	Ft	2 760	1	2 760
Interaction between three parameters	AFt	1 547	2	774
Replication	Residual	2 300	48	48
	Total	74 741	59	

'experimental error', and using the variance ratio test of Snedecor ('F-test'), it can be ascertained whether or not a source of variation has a significant influence on the deviation of the test results. If a higher order interaction proves to be significant, it is not permissible to test lower-order combinations against the residual.

Table 9. Table of Analysis of Variance of the Ultimate Strengths of the Heat-Affected Zone $\sigma_{u_{HAZ}}$ (Tables 5 and 6). Parameters varied were: Alloy A; Filler Metal; F; Thickness, t.

Nature of effect	Source	Sum of squares	Degrees of freedom	Variance estimate
Main parameters	A	642 710	2	321 355
	F	64	1	64
	t	4 002	1	4 002
Interaction between two parameters	AF	21	2	11
	At	2 329	2	1 165
	Ft	0	1	0
Interaction between three parameters	AFt	148	2	74
Replication	Residual	1090	48	23
	Total	650 364	59	

Concerning the ultimate strengths of the weld metal it can be stated that:
* The influence of the combination alloy/filler metal/thickness is very significant. In other words, each combination yields different values for the ultimate strength of the weld metal.
* Mutual comparison of variances reveals a significant influence of the alloy in all cases, whereas the influences of filler metal and thickness are not significant.
* The residual variance is small, i.e. the deviation in results due to the experimental error and no other source of variation, is small.

Proceeding from the above results, in Table 10 characteristic values for the ultimate strength of the weld metal are given for the combinations of alloy/filler metal/thickness that were studied. These values were determined by:

—The residual variance as a measure for the standard deviation (see Table 8: $\sigma = \sqrt{48} = 6·93 \rightarrow \sigma = 7\,N/mm^2$).
—ISO Standard 3207 'Statistical interpretation of data determination of a statistical tolerance interval' for characteristic values with a confidence level of 95%.

Concerning the ultimate strengths of the heat-affected zone the analysis of variance arrived at the following:
* The influence of the combination alloy/filler metal/thickness is hardly significant.
* The influence of the combination alloy/thickness is very significant.
* Mutual comparison of variances shows a very significant influence of the alloy, whereas all other parameters and combinations of parameters are not significant.

Table 10. Characteristic Values (95% Confidence Level) for Ultimate Strength of the Weld Metal σ_{u_w} in N/mm^2

Welding process	Filler metal	Thickness	5083-0	6063-T5	6082-T6	7020-T6
MIG	5356	$t = 4$ mm	237	158	214	238
		$t = 12$ mm	257	200	218	281
	4043	$t = 4$ mm	—	150	194	223
		$t = 12$ mm	—	166	195	214
TIG	5356	$t = 4$ mm	—	163	207	—

Table 11. Characteristic Values (95% Confidence Level) for Ultimate Strength of Heat-Affected Zone $\sigma_{u_{HAZ}}$ in N/mm²

Welding process	Thickness	5083-0	6063-T5	6082-T6	7020-T6
MIG	$t = 4$ mm	285[a]	140	200	370
	$t = 12$ mm	304[a]	145	212	404
TIG	$t = 4$ mm	—	132	158	—

[a] Hardly any influence of heat-input, mean values $\sigma_{u_{HAZ}} = 297$. ($t = 4$ mm) and 316 ($t = 12$ mm) N/mm² (see also Table 5).

- The residual variance again is small (see also the results for the weld metal), so the deviation in results due to the experimental error is small.

In Table 11 characteristic values for the ultimate strengths of the heat-affected zone are given for combinations of alloy and thickness, which are determined in the same way as for the weld metal. (In this case: $\sigma = 5$ N/mm².)

Finally, all the relevant results, both from the literature (see Section 3.2) and the present study, concerning ultimate strengths of the weld metal and heat-affected zone, are presented in Figs. 7–11 for the respective alloys and filler metals. With respect to the results in the above diagrams the following can be observed:

- With a few exceptions comparable results from the literature and from the present study, are in reasonably good agreement.
- For the ultimate strengths of the weld metal:
  —thicker plates yielded higher values, contrary to the literature, but the influence of the thickness was not significant;
  —the TIG results for thin plates agree very well with comparable results for MIG.
- For the ultimate strengths of the heat-affected zone:
  —for 'normal' choices of welding parameters and welding process similar values were found in the present study and the literature. In 'abnormal' circumstances (see Table 2: TIG + $t = 20$ mm) considerably lower values may be found especially for 6082-T6;
  —the results of series 4 (see Section 4.1) in Figs 8 and 9, which correspond to the numbers 9 and 10 in the plots, appeared to be within the scatter band of the results of the reference programme (series 1, 2 and 3).

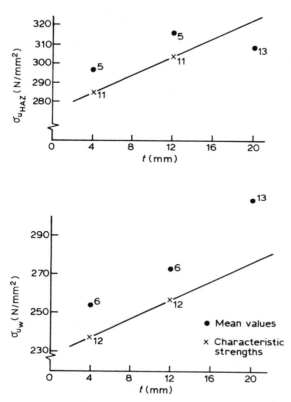

Fig. 7. Failure strengths 5083−0 and F+5356. The numbers in the diagrams correspond to: 5, 6, 11 and 12, TNO research;[13] 13, IIW research.[7]

4.5 Conclusions relating to mechanical properties

The following conclusions can be drawn from the results relating to mechanical properties as discussed in this section:

- The results from this study—both for the weld metal and the heat-affected zone—agree reasonably well with comparable results from the literature.
- Based on the statistical analysis of the results, it was pointed out that:
  —the ultimate strength of the weld metal depends on the combination alloy/filler metal/plate thickness (weld type). However, the

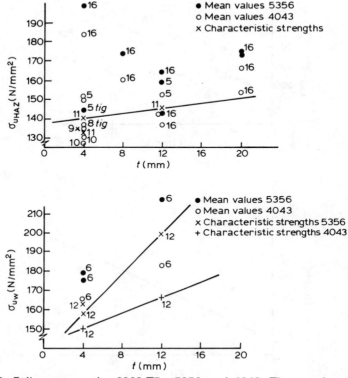

Fig. 8. Failure strengths 6063-T5 – 5356 and 4043. The numbers in the diagrams correspond to: 5, 6, 8, 9, 10, 11 and 12, TNO;[13] 16, Pechiney.[10]

welding process has no significant influence. (TIG and MIG yield similar results in the case of relatively thin plates);
—the ultimate strength of the heat-affected zone depends on the combination alloy/time–temperature relationship in the plate material. The TIG-process yields less good results than the MIG-process;
—both for the ultimate strengths of the weld metal and for the heat-affected zone characteristic values have been determined which provide a suitable basis for the design of welded connections in aluminium alloys.
• The results of the additional programme (series 4) proved to be within the scatter band of the results of the reference programme.

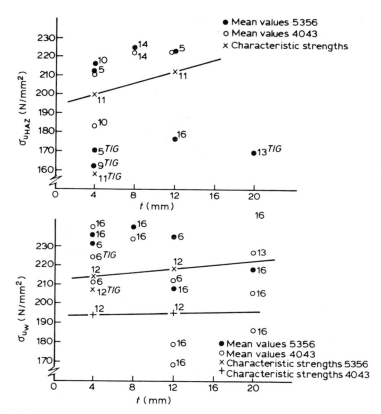

Fig. 9. Failure strengths 6082-T6 + 5356 and 4043. The numbers in the diagrams correspond to: 5, 6, 9, 10, 11 and 12, TNO;[13] 13, IIW (TIG results);[7] 14, FMPA (Germany);[8] 16, Pechiney.[10]

- The results of the hardness measurements do not suggest a better approximation of the dimensions of the heat-affected zones than the '1-in' rule.

5 EXPERIMENTAL RESEARCH ON FILLET WELDS

5.1 Choice of parameters

In view of the results of the literature study concerning fillet welds, and the investigation of mechanical properties, an experimental

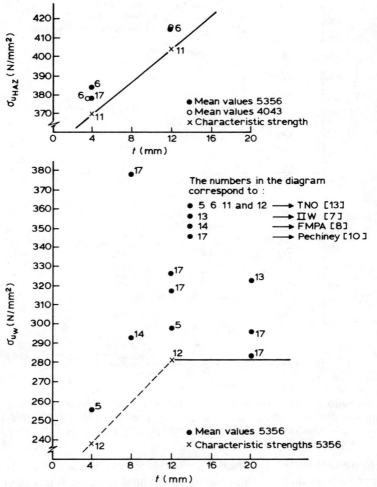

Fig. 10. Failure strengths 7020-T6 + 5356 and 4043.

research study on fillet welds was carried out with the following parameters:

- alloys and filler metals identical to the research on mechanical properties (see Section 4.1);
- two types of connections, see Fig. 12;

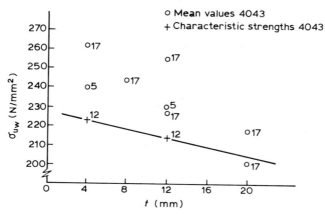

Fig. 11. Failure strengths 7020-T6 + 4043. The numbers in the diagram correspond to: 5, 12, TNO;[13] 17, Pechiney.[10]

- 1 plate thickness, namely $t = 12$ mm;
- 1 throat thickness fillet weld, namely $a = 5$ mm;
- 1 welding process, namely MIG (spray-arc).

In this programme the following were not varied:

- the welder: all plates were welded by the same, qualified welder;
- the welding position: only the flat position (2F) was applied;
- the welding parameters: the choice of welding parameters was determined by the weld quality which was prescribed in the welding procedure specification.

Under the test programme on mechanical properties, the combinations of parameters as shown in Tables 12 and 13 were investigated in the experimental research on fillet welds.

5.2 Welding test specimens

Welding of the test specimens was carried out in accordance with a detailed welding procedure specification concerning edge preparation, cleaning, tacking, welding position, i.e. 2 F, welding sequence, qualification of welding procedure and welder, and recommendations for the welding parameters to be used. The qualification was based on visual inspection of the fillet welds (throat a, equal leg lengths, undercut, overlap and cracks), except for the root penetration which

Table 12. Results for Ultimate Strengths of Fillet Welds Loaded $\perp$ to the Weld Axis ($\tau_{uw\perp}$)

Specimen						Rupture section welds				
Type	Filler metal	Alloy	F_u (kN)	b (mm)	t (mm)	σ (N/mm²)	a_1 (mm)	a_2 (mm)	l (mm)	$\sum al$ (mm²)	$\tau_{uw\perp}$ (N/mm²)	σ_c/β (N/mm²)
			(1)	(2)	(3)	(4)	(5)	(6)	(7)	(8)	(9)	(10)
		5083	143·5	51·0	13·0	216	8·4	8·4	51·0	857	167	236
			144·0	50·0		222	9·2	9·1	50·0	915	157	222
			143·4	50·5		218	8·3	8·9	50·5	869	165	233
			150·0	50·5		228	8·6	7·9	50·5	833	180	255
			149·9	51·2		224	9·1	8·4	51·2	896	166	235
		6063	107·5	49·2	12·3	179	8·1	8·2	49·2	802	134	190
			114·0	50·7		184	8·1	8·1	50·7	821	139	197
			107·2	50·8		178	8·0	8·5	49·4	817	131	185
			106·0	51·0		170	8·9	8·9	51·0	801	118	165
			114·0	50·5		184	8·0	8·6	50·7	906	135	191
	5356	6082	147·5	49·7	12·0	247	—	—	—	—	>175	>248
			144·0	49·0		247	—	—	—	—	>173	>247
			145·2	50·2		241	8·3	8·4	50·2	838	173	247
			139·5	48·9		238	9·0	8·0	48·9	831	168	238
			149·5	50·9		245	9·3	8·0	50·9	881	170	240

		145.8	50.2		238	9.2	8.2	50.5	873	167	236
		146.2	51.6		232	8.0	9.3	51.6	893	164	232
	7020	146.5	50.3	12.2	239	8.8	8.4	50.3	865	169	239
		138.1	49.5		229	7.8	8.2	49.5	792	174	246
		134.5	49.6		222	7.2	8.7	49.6	789	171	242
		93.8	50.2		153	8.2	7.7	50.2	798	118	167
		95.5	50.2		156	8.1	7.7	50.2	793	120	170
	6063	96.5	51.6	12.2	153	7.7	7.5	51.6	784	123	174
		90.3	51.5		144	8.0	7.8	51.5	814	111	157
		95.2	50.7		154	8.2	7.2	50.7	781	122	173
		106.2	50.4		172	8.8	7.9	50.4	842	126	178
		108.9	49.2		181	9.0	8.3	49.2	851	128	181
4043	6082	114.0	50.8	12.2	184	8.7	8.0	50.8	848	134	190
		104.5	50.3		170	8.4	7.9	50.3	820	127	180
		107.4	52.4		168	8.2	8.1	52.4	854	126	178
		133.0	50.2		217	9.0	8.0	50.2	853	156	221
		136.5	50.4		222	9.0	8.0	50.4	857	159	225
	7020	130.5	50.0	12.2	214	9.0	8.5	50.5	875	149	211
		129.0	50.5		209	8.6	8.6	50.5	869	149	211
		131.0	51.5		209	8.0	8.6	51.5	855	153	216

Table 13. Results for Ultimate Strengths of Fillet Welds Loaded ∥ to the Weld Axis ($\tau_{uw\parallel}$)

Specimen						Rupture section welds				
Type	Filler metal	Alloy	F_u (kN)	b (mm)	t (mm)	σ (N/mm²)	a_1t/ma_4 (mm)	l_1t/ml_4 (mm)	$\sum al$ (mm²)	$\tau_{uw\parallel}$ (N/mm²)	σ_c/β (N/mm²)
			(1)	(2)	(3)	(4)	(5)	(6)	(7)	(8)	(9)
		5083	254·0	69·9	12·4	147	8·3	51·0	1693	150	260
			239·8	69·8		139	7·9	51·0	1612	149	258
			257·0	69·9		148	8·5	51·0	1734	148	256
			235·0	70·0		135	7·5	51·0	1530	154	267
			234·0	69·9		135	7·7	50·0	1540	152	263
	5356	6063	172·8	70·0	12·2	101	7·8	49·0	1529	113	196
			172·0	70·0		101	7·5	50·0	1500	115	199
			173·0	69·8		102	7·5	50·0	1500	115	199
			176·8	69·8		104	8·4	49·0	1646	107	185
			174·0	69·9		102	8·0	50·0	1600	109	189
		6082	188·0	69·9	12·2	110	8·2	51·0	1673	112	194
			189·0	70·0		111	7·7	51·0	1571	120	208
			202·0	69·6		118	8·1	51·0	1652	122	211

Alloy								
	188.0	70.0		110	7.4	51.0	1510	125	217
	184.1	70.0		108	8.3	51.0	1693	109	189
7020	248.2	69.9		146	7.9	51.0	1612	154	267
	250.5	70.0		147	8.1	51.0	1652	152	263
	229.8	69.6	12.2	135	7.5	50.0	1500	153	265
	233.5	69.6		137	7.9	52.0	1643	142	246
	247.0	69.3		146	8.1	51.0	1652	150	260
6063	159.0	69.5		94	7.2	51.0	1469	108	187
	140.0	69.4		83	7.3	41.0	1197	117	203
	146.2	69.4	12.2	86	6.5	51.0	1326	110	191
	150.0	69.7		88	6.3	52.0	1310	114	197
	140.0	69.7		82	6.6	51.0	1346	104	180
6082	172.4	69.9		101	7.2	51.8	1492	116	201
	177.5	70.0		104	7.0	52.0	1456	122	211
	172.9	69.6	12.2	102	6.8	51.5	1401	123	213
	187.2	69.5		110	7.7	51.3	1580	118	204
	183.0	69.7		108	7.1	52.5	1491	123	213
7020	191.0	69.5		113	6.7	51.3	1375	139	241
	191.0	69.0		113	7.1	51.0	1448	132	229
	195.0	69.2	12.2	115	7.0	50.4	1411	138	239
	217.5	69.7		128	7.4	50.7	1501	145	251
	189.0	69.0		112	7.1	50.8	1443	131	227

4043

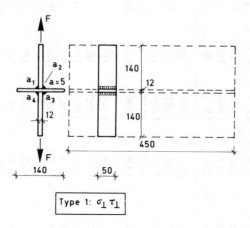

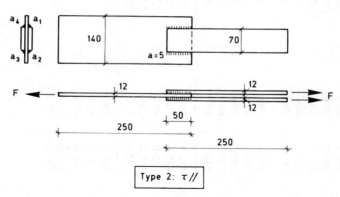

Fig. 12. Test specimens for fillet welds, types 1 and 2; dimensions in mm.

was tested destructively. Ensuring sufficient root penetration appeared to be most difficult of all criteria to satisfy. In Ref. 15 it was concluded that sufficient root penetration is achieved when the arc voltage and the current are as high as possible, namely just below the level where burning of weld metal occurs.

The specimens used in this programme are shown in Fig. 12. With the specimens of type 1 four welds (a_1 up to a_4) of 450 mm length were laid down. From this length five test specimens, each 50 mm in length, were taken (at both ends a length of 100 mm was not used). In the case of type 2 each test specimen was welded separately, namely five test specimens for each combination of alloy and filler metal. In the welding of these specimens, 'run-on, run-off' plates were used.

5.3 Testing

The specimens of types 1 and 2 were loaded as shown in Fig. 12. With these tests only the failure strength of the specimens was recorded. Before testing, the dimensions of the throat section, length and throat thickness (a) were measured. After testing, the dimensions of the rupture section and the angle (ϕ), as indicated in Fig. 13, were measured.

The throat thickness (a) was measured from the outside. In addition, for each combination of alloy and filler metal, one macro-etch was made in order to measure the effective throat thickness (a_{eff}) and the penetration (p), see Fig. 13. In Tables 14–16 the results of these measurements are summarized, i.e. only average values are given. A period of about 6 weeks elapsed between welding and testing.

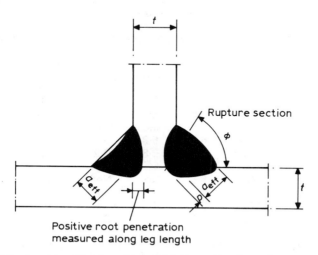

Fig. 13. Fillet weld: effective throat thickness (a_{eff}), and penetration (p) both measured in the throat section; positive root penetration. The situation and the angle ϕ of the rupture section are also indicated.

Table 14. Test Specimen Type 1. Average Values Dimensions of Throat Section And Rupture Section

Test specimen			Throat section			Rupture section			
Type	Filler Metal	Alloy	a (mm)	a_{eff} (mm)	p (mm)	a_1 (mm)	a_2 (mm)	ϕ_1	ϕ_2
1	5356	5083	5·9	6·1	2·0	8·7	8·5	55°	65°
		6063	5·5	6·0	2·3	8·2	8·5	55°	65°
		6082	5·5	6·0	2·6	8·3	7·7	60°	70°
		7020	6·5	6·4	2·3	8·2	8·6	55°	70°
	4043	6063	5·5	5·8	2·1	8·0	7·6	60°	70°
		6082	5·5	6·1	1·6	8·6	8·0	55°	70°
		7020	5·8	5·8	4·0	8·7	8·3	45°	60°

Note: Throat section—average values of four welds (each weld length 450 mm).
Rupture section—average values of five tests, rupture two welds a_1a_2 or a_3a_4 (each weld lengths 50 mm, see Fig. 12).

Table 15. Test Specimen Type 2. Average Values Dimensions of Throat Section and Rupture Section

Test specimen			Throat section			Rupture section	
Type	Filler metal	Alloy	a (mm)	a_{eff} (mm)	p (mm)	a_1 up to a_4 (mm)	ϕ_1 up to ϕ_4
2	5356	5083	5·8	5·4	3·3	8·0	55°
		6063	6·5	Not meas.	Not meas.	7·8	57°
		6082	6·0	5·8	3·2	7·9	58°
		7020	5·9	5·8	2·2	7·9	57°
	4043	6063	5·5	Not meas.	Not meas.	6·8	49°
		6082	5·3	4·9	3·8	7·0	53°
		7020	5·2	4·8	3·8	7·1	49°

Note: Throat section—average values of five test specimens; for each test specimen four welds of 50 mm length, see Fig. 12.
Rupture section—see above.

Table 16. Test Specimen Type 1, Additional Tests. Average Values Dimensions of Throat Section and Rupture Section

Test specimen			Throat section			Rupture section			
Type	Filler metal	Alloy	a (mm)	a_{eff} (mm)	p (mm)	a_1 (mm)	a_2 (mm)	ϕ_1	ϕ_2
1	5356	5083	5·5	5·8	2·4	8·0	7·8	55°	70°
		6082	5·5	5·4	2·1	7·8	7·6	55°	70°
		7020	6·3	5·9	2·5	8·1	8·1	57°	65°
	4043	6082	5·6	5·9	4·6	8·9	8·3	55°	60°

Note: Throat section—average values of four welds (each weld length 450 mm).
Rupture section—average values of five tests, rupture two welds, a_1a_2 or a_3a_4 (each weld length 50 mm, see Fig. 12)

5.4 Test results for fillet welds

To determine the ultimate strength of the fillet welds, five tensile tests per type of connection were carried out, for every combination of parameters investigated.

The results of the tensile tests on the two types of connections are given in Tables 12, 13 and 17: in Tables 12 and 17 results of connection type 1, loaded perpendicularly to the weld axis, and in Table 13 results of connection type 2, loaded parallel to the weld axis.

The ultimate strength of the respective welds was determined by dividing the failure load (F_u) by the total area of the rupture section ($\sum al$). For the connection of type 2 this yields the ultimate shear strength $\tau_{uw\parallel}$ and for the connection of type 1 the ultimate shear strength $\tau_{uw\perp}$. However, the latter is not correct since failure is due to a combination of tensile and shear actions on the rupture section, so this is not a 'real' shear strength. In order to make a comparison with similar results from the literature study (see Section 3.3.2), for the connection of type 1 the value of $\tau_{uw\perp}$ was calculated.

The results of *additional* tests on connections of type 1 are given in Table 17. After the specimens for the reference tests had been welded (Table 12), additional specimens were welded with similar welding parameters except for a higher travel speed. This yielded higher results (up to 15%) as can be inferred by comparing corresponding results of Tables 12 and 17.

Besides the reference programme described above, an additional

Table 17. Results for Ultimate Strength of Fillet Welds (Additional Tests), Loaded Perpendicularly to the Weld Axis ($\tau_{uw_\perp}$)

Test specimen			Rupture section					$\dfrac{\sigma_c}{\beta}$
			F_u (kN)	a_1 (mm)	a_2 (mm)	l (mm)	$\sum al$ (mm)	$\tau_{uw_\perp}$ (N/mm^2)	(N/mm^2)
Type	Filler metal	Alloy	(1)	(2)	(3)	(4)	(5)	(6)	(7)
1	5356	5083	146·4	7·5	8·2	50·3	790	185	262
			147·0	7·5	7·5	51·0	765	192	272
			163·5	8·6	7·6	52·0	842	194	274
			158·5	8·2	7·9	50·5	813	195	276
			156·5	8·4	7·7	50·0	805	194	274
			154·2	8·1	8·5	49·1	815	189	267
			156·5	8·6	7·6	50·5	818	191	270
			154·5	8·1	7·8	50·3	800	193	273
		7020	146·5	8·0	8·4	49·8	817	179	253
			155·2	8·7	7·5	50·3	815	190	269
			154·9	8·1	8·8	49·3	833	186	263
			157·0	8·0	8·3	50·0	815	193	273
			150·4	7·6	8·1	49·4	776	194	274
			155·2	7·7	7·7	50·3	775	200	283
	4043	6082	118·0	8·7	8·6	49·8	862	137	194
			134·5	9·2	8·0	52·2	898	150	212
			129·5	7·8	8·9	50·5	843	154	218
			133·0	9·0	7·9	51·0	862	152	218
			135·7	9·7	7·9	51·5	906	150	212

programme was carried out by the other firms, similar to the one described in Section 4.1, applying different alloys, filler metals, plate thicknesses, throat thicknesses and welding processes (MIG and TIG). All the results are discussed in Ref. 16, except the results of the latter additional programme, which have been summarized in Ref. 21.

5.5 Evaluation of reslts

5.5.1 Weld areas

Although the welding was not a parameter in this test programme (see Section 5.1) the following can be observed regarding the weld areas (see also Tables 14–16):

- For the throat sections:
  —A throat thickness of 5 mm was achieved in almost all cases.
  —The effective throat thickness (a_{eff}) is satisfactorily estimated by the throat thickness measured from the outside (a). This is the

case as long as the welds possess equal leg length, the welds are not too convex (Fig. 13: right fillet weld), and positive root penetration is ensured (required according to the welding procedure specification).
—With positive root penetration the values of the penetration p, measured along the throat section of the weld, see Fig. 13, were found to be rather high.
- For the rupture sections:
—The thicknesses a_1 and a_2 (type 1) and a_1-a_4 (type 2) are significantly higher than the values of the corresponding throat sections, which is due to the considerable penetration observed. Although the situation and the orientation (angle ϕ) of the rupture section differs from the throat section (see Fig. 13), the thickness of the rupture section can be satisfactorily estimated from the effective throat thickness plus the penetration p (a_1, a_2, respectively a_1 up to $a_4 = a_{\text{eff}} + p$).
—The variation of the rupture thicknesses was small; in general a_1, a_2, and $a_1-a_4 \simeq 8\cdot 0 \pm 0\cdot 5$ mm, except for the connection of type 2 combined with filler metal 4043 (see Table 15): a_1 up to $a_4 \simeq 7\cdot 0 \pm 0\cdot 5$ mm. Differences between the type 1 reference tests and the additional tests with respect to the thickness of the rupture sections were not significant.
—The values of the angle ϕ varied between 40° and 75°. For the connection of type 1: mean values $\phi \simeq 60°$, for type 2: $\phi = 55°$; in combination with filler metal 5356 higher values were found than with filler metal 4043.

The ultimate strength of the fillet welds was based on the area of the rupture section.

5.5.2 Ultimate strengths

The results of the ultimate strengths of the fillet welds investigated were evaluated statistically by means of analyses of variances, as was also carried out in the research programme on mechanical properties (see Section 4.4). In Ref. 16 such an analysis was performed using the results of Tables 12 and 13, i.e. the reference tests. The analysis of variance is given in Table 18. With this analysis it was ascertained that:

- The influence of the combination alloy/filler metal/type of specimen (load direction) is very significant. Each combination yields different values for the ultimate strength of the fillet welds.

Table 18. Table of Analysis of Variance of the Ultimate Strength of Fillet Welds (Results).
Parameters varied: A, alloy; F, filler metal; t, type of specimen.

Nature of effect	Source	Sum of squares	Degrees of freedom	Variance estimate
Main parameters	A	11 738	2	5 869
	F	2 898	1	2 898
	t	6 469	1	6 469
Interaction between 2 parameters	AF	375	2	188
	At	729	2	365
	Ft	1 420	1	1 420
Interaction between 3 parameters	AFt	1 527	2	764
Replication	Residual	1 191	48	25
	Total	26 347	59	

- The residual variance is small, i.e. the deviation in results due to the experimental error, and no other source of variation, is small.

Besides the analyses of variances as described, the results of tests on fillet welds were compared with those of the literature study (see Section 3.3.2). In Figs 14–17 all the results for combinations of alloy and filler metal are given, i.e. mean values for the ultimate shear strengths of fillet welds, loaded perpendicularly ($\tau_{uw_\perp}$) or parallel ($\tau_{uw_\parallel}$) to the axis of the weld for different throat thicknesses (a).

- Generally the results of FMPA[8] and TNO (both reference and additional results) are in very good agreement, whereas the results of IIW[7] and Pechiney[10] do not agree very well with the other results, especially for $\tau_{uw_\perp}$; in addition, these results show considerable scatter. The main reason for this deviation is that the results of IIW and Pechiney are based on the throat section (measured from outside), whereas the results of FMPA and TNO are based on the rupture section. Besides, the type of specimen used by IIW and Pechiney is not very well suited to the determination of ultimate strengths of fillet welds.
- The TIG results, only the IIW test,[7] are below the corresponding MIG results. (Most values in the diagrams are MIG results, except the values with a TIG label.)

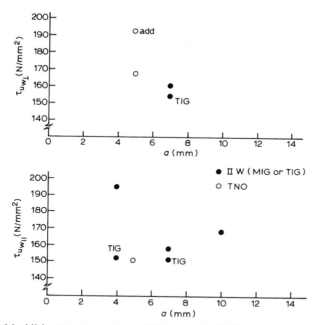

Fig. 14. Ultimate strengths of fillet welds 5083 − 0 and F + 5356.

Summarizing, it is believed that the TNO results discussed are a reliable lower limit of the ultimate strengths of fillet welds. The literature results are mostly higher, sometimes lower, but the differences can be easily explained, as described before.

5.5.3 Design of fillet welds

As already stated (see Section 3.3.2 and Ref. 11), the actual strength of a fillet can be approximated very well by applying the β-formula, i.e.

$$\sigma_c = \sqrt{\sigma_\perp^2 + 3(\tau_\perp^2 + \tau_\parallel^2)} \leq \sigma_d$$

where

σ_c	= comparison stress,
$\sigma_\perp$, $\tau_\perp$ and $\tau_\parallel$	= stresses in throat section according to Fig. 2,
σ_d	= design strength,
β	= factor; the value of β is determined by differences in strength and ductility of fillet weld metal and butt weld metal, and the difference between the value 3 in the formula and the 'exact' value 2·6.

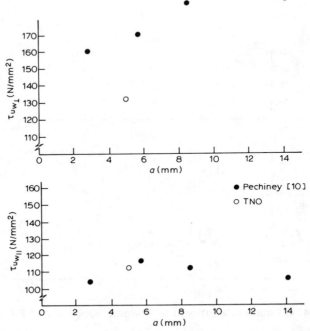

Fig. 15a. Ultimate strengths of fillet welds 6063-T5 + 5356.

For applying this formula, the values of β and σ_d have to be known. The design strength σ_d should be based on the ultimate strength of the weld metal σ_{uw} instead of $\sigma_{0.2_w}$ since the design rules are based on (ultimate) limit states for which the yield strength of the weld ($\sigma_{0.2w}$) is of no significance. For this purpose the results of the experimental research on mechanical properties (see Section 4) can be used, i.e.

$$\sigma_d = \frac{\sigma_{uw}}{\gamma_u}$$

where

σ_{uw} = characteristic value for the ultimate strength of weld metal dependent on alloy and filler metal,

γ_u = member resistance factor.

Similarly to the determination of β in the case of steel structures (see Ref. 17) the value of β has been calculated by comparing the ultimate

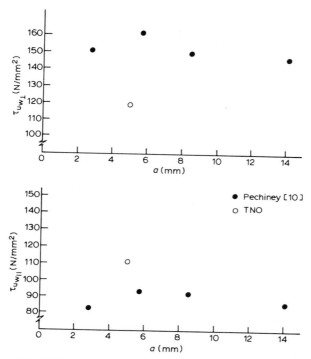

Fig. 15b. Ultimate strengths of fillet welds 6063-T5 + 4043.

strengths of fillet welds to the ultimate strengths of the corresponding weld metal.

First, the values of σ_c/β have been determined for both types of specimen as follows:

Specimen type 1: $\sigma_\perp = \tau_\perp = \frac{1}{2}\sqrt{2}\tau_{uw_\perp} \rightarrow \frac{\sigma_c}{\beta} = \tau_{uw_\perp}\sqrt{2}$

(see Tables 12 and 17; column 7)

Specimen type 2: $\tau_\| = \tau_{uw\|} \rightarrow \frac{\sigma_c}{\beta} = \tau_{uw\|}\sqrt{3}$ (see Table 13, column 6)

Note: the values of $\sigma_\perp$, $\tau_\perp$ and $\tau_\|$ have been determined assuming that the throat and rupture section coincide, which simplifies the calculation and is permissible, as demonstrated in Section 5.5.1.

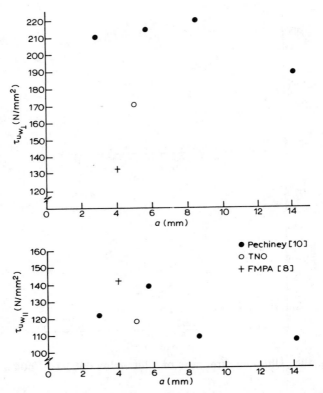

Fig. 16a. Ultimate strengths of fillet welds 6082-T6 + 5356.

Second, the average values of σ_c/β have been compared with the average values of the ultimate strengths of the corresponding weld metal σ_{uw}. The latter values were taken from Figs 7–11 (see also Section 4.4), i.e. the values were calculated by linear interpolation for a thickness $t = 8$ mm, which almost equals the thickness of the rupture section of the fillet welds.

Comparing average values was preferred because:

— the number of test results was restricted due to the number of combinations investigated;
— the results of σ_c/β do not have a normal distribution.

The results used for this comparison and the results for β are given in

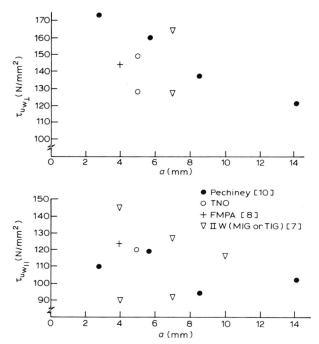

Fig. 16b. Ultimate strengths of fillet welds 6082-T6 + 4043.

Table 19. The results for β have been derived as follows:

$$\sigma_c \leq \sigma_{uw} \quad \text{or} \quad \frac{\sigma_c}{\beta} \cdot \beta \leq \sigma_{uw} \quad \text{or} \quad \beta = \frac{\sigma_{uw}}{\sigma_c/\beta}$$

Thus, in Table 19 the results of column (1) have to be divided by the results of column (2), (3) or (4), which yields the values for β as given in columns (5) and (6). For design purposes a value $\beta = 1 \cdot 0$ can be applied based on the mean values of β in column (6) because:

- The higher the value of β, the lower the strength of the fillet weld. An average value of β has been calculated because:
- —the lowest value of β is determined by the (relatively) low results of column (2);
- —in the design the throat section is used instead of the rupture section;
- —in applying the β-formula, the design strength σ_d is used instead of σ_{uw}.

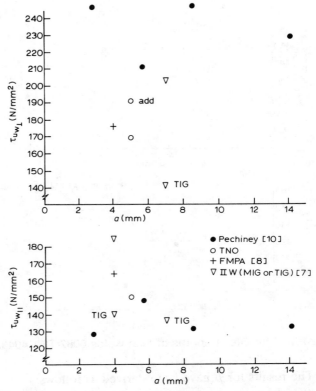

Fig. 17a. Ultimate strengths of fillet welds 7020-T6 + 5356.

5.6 Conclusions relating to fillet welds

From the results on fillet welds as discussed in this section the following can be concluded:

- The thickness of the rupture section of a fillet weld can be significantly higher than the throat thickness, this being due to positive root penetration. This penetration has to be ensured in the qualification procedure.
- The location and the orientation of the rupture section justify the assumption that the rupture and throat section of the fillet weld coincide.

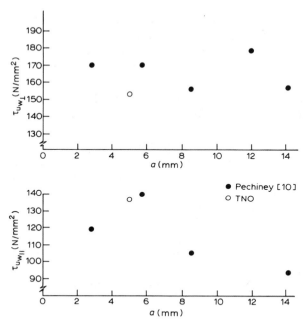

Fig. 17b. Ultimate strengths of fillet welds 7020-T6 + 4043.

- For the ultimate strength of fillet welds it is found that:
  —The influence of the combination alloy/filler metal/load direction (type of specimen) is very significant, which was demonstrated by the analysis of variances of the results.

Table 19. β-Values for the Design of Fillet Welds

Alloy	Filler metal	σ_{uw} (N/mm^2)	σ_c/β (N/mm^2)			β	
			Type 1 (Table 12)	Type 1 (Table 17)	Type 2 (Table 13)	Min–max	Mean values
		(1)	(2)	(3)	(4)	(5)	(6)
5083-0	5356	264	236	272	261	0·97–1·12	1·03
6063-T5	5356	196	191	203	194	0·97–1·03	0·99
	4043	175	168	—	192	0·91–1·04	0·98
6082-T6	5356	233	244	—	204	0·95–1·14	1·05
	4043	212	181	211	208	1·00–1·17	1·06
7020-T6	5356	276	239	272	260	1·01–1·15	1·07
	4043	235	217	—	237	0·99–1·08	1·04

—The load direction influences the ultimate strength, although to a much lesser extent than sometimes asserted, see Section 3.3.2.
—MIG results appear to be better than TIG results (from literature), which is possibly due to better penetration achieved with MIG welding.
—The results of this experimental research on fillet welds are found to provide a reliable lower limit for the ultimate strength of fillet welds.
- For the design of fillet welds it is found that:
  —The β-formula, which can be similarly applied to steel structures, gives the best approximation of the strength of a fillet weld compared with other methods (see Refs 1 and 2).
  —If, applying the β-formula, the design is based on the characteristic strength of the weld metal and the dimensions of the throat section, and a value $\beta = 1 \cdot 0$ (see Table 19) is adopted, a design strength is obtained which is a reliable lower limit of the actual ultimate strength.
  —For simple connections simple design formulas derived from the β-formula can be applied.

6 EXPERIMENTAL AND THEORETICAL RESEARCH ON WELDED CONNECTIONS

6.1 General

Welded connections of rectangular hollow sections, in which the brace section (vertical) was connected to the chord section (horizontal) by means of a fillet weld along the entire perimeter, were investigated in this study. These connections were chosen because:

- This type of connection yields a very unequal stress and strain distribution which is well suited for investigating the influence of the mechanical properties on the structural behaviour of the connection, see Section 1.
- The type of connection chosen is relatively simple and for steel structures much research has been devoted to similar connections.

Within the scope of the experimental and theoretical research described in this section, we have attempted, proceeding from the results obtained for the mechanical properties and fillet welds, to ascertain the structural behaviour of welded connections in aluminium alloy structures.

6.2 Choice of parameters

Referring to the choice of parameters in the foregoing studies (see Sections 4.1 and 5.1), the following parameters were chosen in this research:

- two alloys, namely 6063-T5 and 7020-T6;
- one filler metal, namely 5356;
- two combinations of sections, namely $100 \times 100 \times 4 + 80 \times 80 \times 4$ and $100 \times 100 \times 4 + 50 \times 50 \times 4$;
- two types of connection, two different loadings (see Fig. 18);
- one throat thickness fillet weld, namely $a = 4$ mm;
- one welding process, namely MIG (spray-arc).

An important reason for this choice of alloys, filler metal and weld thickness was that, starting from the results of the foregoing studies, the strength of the weld is higher than the strength of the heat-affected zone for 6063-T5, while for 7020-T6 the reverse is true.

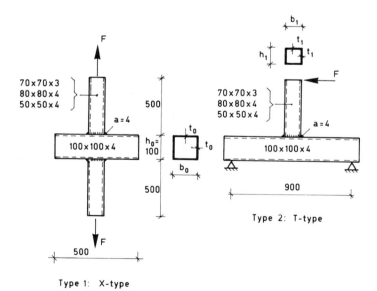

Fig. 18. Types of hollow sections investigated; dimensions in mm; loadings.

Table 20. X-Type Specimens. Results for Failure Load (F_u), Ultimate Displacements (δ_{1u} resp. δ_{2u}) and Failure Modes

Alloy	Combination	Test	F (kN)	δ_{1u} (mm)	δ_{2u} (mm)	Failure mode
6063-T5	100 × 100 × 4 + 80 × 80 × 4	A	81·5	4·5	7·0	Cracking at the toe of the weld of the bracings at front and back respectively Cracking starts at the corners
		B	89·6	6·2	8·3	
		C	96·0	—	—	
		D	104·0	7·6	10·4	
		E	103·6	8·0	11·1	
	100 × 100 × 4 + 50 × 50 × 4	A	45·7	24·0	25·6	Large deformations of chord section (chord faces) and subsequently tearing of chord side walls
		B	48·0	25·3	29·4	
		C	61·3	28·0	30·0	
		D	49·0	27·5	29·8	
		E	—	—	—	
7020-T6	100 × 100 × 4 + 80 × 80 × 4	A	105·8	2·3	3·1	→ See above, cracking at toe of bracing weld
	100 × 100 × 4 + 50 × 50 × 4	A	42·8	12·0	12·0	→ See above, cracking at toe of bracing weld
		B	49·0	12·0	12·0	→ Cracking of welds at front and back and cracking of chord section at front and back respectively
	100 × 100 × 4 + 70 + 70 × 3	A	63·5	—	—	→ See above, cracking at toe of bracing weld

As in the preceding investigations, the following were not varied: the welder, the welding position (2F) and the welding parameters (determined by the weld quality as prescribed in the welding procedure specification).

In connection with the foregoing test programmes on mechanical properties and fillet welds, the combinations of parameters shown in Tables 20 and 21 were investigated.

6.3 Welding of hollow sections

Welding of the test specimens was again carried out according to a detailed welding procedure specification, which was very similar to that described in Section 5.2. The dimensions of the test specimens and the throat thicknesses are given in Fig. 18. Actually the throat thickness (a), measured from the outside, varied between 4 and 6 mm.

Table 21. T-Type Specimens. Results for Failure Load (F_u), Ultimate Displacements (δ_{1u} resp. δ_{2u}), Failure Moment (M_u), Ultimate Rotation/ϕ_u) and Failure Modes

Alloy	Combination	Test	F_u (kN)	δ_{1u} (mm)	δ_{2u} (mm)	M_u (kN/m)	δ_u (rad)	Failure mode
6063-T5	100 × 100 × 4 + 80 × 80 × 4	A	6·9	6·5	105	3·35	216 × 10⁻³	→ Cracking at the toe of the welds at the corners of the bracing
		B	7·1	7·5	120	3·44	247 × 10⁻³	→ Cracking of chord sidewalls
		C	6·3	8·1	120	3·06	247 × 10⁻³	→ Similar to Test A above
		D	4·8	5·9	90	2·86	186 × 10⁻³	→ Similar to Test B above
	100 × 100 × 4 + 50 × 50 × 4	A	5·6	—	—	2·72	—	Cracking at toe of the welds of the bracing
		B	2·7	7·0	140	1·31	289 × 10⁻³	
		C	4·7	9·0	180	2·28	370 × 10⁻³	Cracking chord face after large deformations
		D	3·9	8·5	170	1·89	350 × 10⁻³	
7020-T6	100 × 100 × 4 + 80 × 80 × 4	A	6·9	6·5	94	3·35	194 × 10⁻³	→ Cracking at the toe of the welds at the corners of the bracing
	100 × 100 × 4 + 50 × 50 × 4	A	3·6	8·5	160	1·75	330 × 10⁻³	
		B	3·4	7·2	146	1·65	301 × 10⁻³	
	100 × 100 × 4 + 70 × 70 × 3	A	5·6	—	118	2·72	243 × 10⁻³	Cracking chord face after large deformations

A detailed description of the welding qualification and test specimens is given in Ref. 18. From a structural point of view it is important that the tacking and the start and stop sites of the welds were chosen in the middle of the side walls of the brace section.

6.4 Testing

The test specimens, type 1 (X-type) resp. type 2 (T-type) were loaded as indicated in Fig. 18. The following data were recorded in the tests:

- the loads (F) up to failure of the connection;
- the displacements (δ) as shown in Fig. 19;
- the strains measured by strain gauges as shown in Fig. 20 for an X-type specimen;
- strains measured according to the moiré-method (see also Section 4.3, and Figs 5 and 6).

6.5 Theoretical research

For the theoretical part of this study the TNO-IBBC finite element programme DIANA[19] was used to simulate the behaviour up to failure of four X-type specimens. With these simulations physical non-linearity ('true' stress–strain diagrams for parent metal, heat-affected zone and weld metal) and geometrical non-linearity (large rotations and displacements) were taken into account. An example of an element distribution is shown in Fig. 21.

6.6 Experimental and theoretical results

The results of tests on the X-type specimens are given in Table 20. A typical load–displacement diagram is shown in Fig. 22, and the corresponding theoretical results are also plotted in this diagram.

For this test specimen the results of the strain gauge measurements are given in Fig. 23. For load step 9 the results of all 22 strain gauges have been compared with the theoretical results for this X-type specimen, see Fig. 24.

The results of the tests on the T-type specimens are summarized in Table 21. A complete survey of all the results is given in Ref. 20.

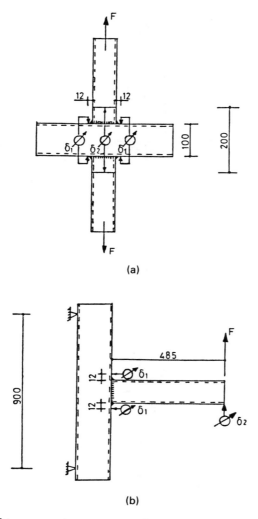

Fig. 19. (a) X-type specimen, measuring displacements δ_1 and δ_2; (b) T-type specimen, measuring displacements δ_1 and δ_2; dimensions in mm.

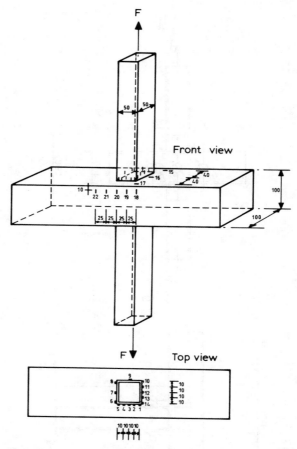

Fig. 20. X-type specimen; dimensions in mm; position and numbering of strain gauges.

6.7 Evaluation of results and conclusions

From the results on welded connections in hollow sections the following was inferred:

- For both types of connection investigated, namely X-type tensile loading and T-type moment plus shear loading, there is no difference in the ultimate strength of the connections for the alloys 6063-T5 and 7020-T6 respectively.

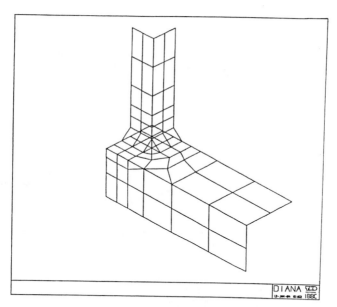

Fig. 21. X-type connection; element distribution combination 100 × 100 × 4 + 50 × 50 × 4.

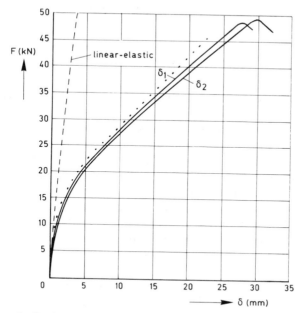

Fig. 22. Load–displacement diagram ($F-\delta$) specimen X-type, 6063-T5, 100 × 100 × 4 + 50 × 50 × 4 mm. · · · · ·, results of computer program; ———, results of tests.

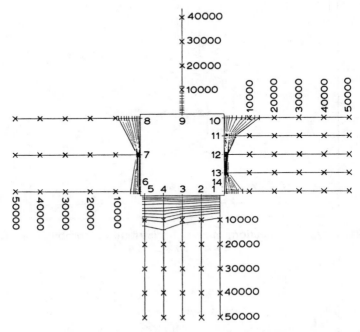

Fig. 23. X-type specimen, alloy 6063-T5, combination 100 × 100 × 4 + 50 × 50 × 4 mm. Strain values gauges 1–14 (see Fig. 20); values in μm/m.

- With the X-type connections the deformations for the alloy 6063-T5 are two or three times higher than those for 7020-T6.
- With the T-type connections the differences in deformations between the alloys 6063-T5 and 7020-T6 are negligible, since the geometry of the connection dominates other influences.
- For both types of connection the same failure modes have been observed for both alloys.
  With the X-type connections there occurs:
  —cracking at the toe of the weld of the bracings (front resp. back side);
  —large deformations of the chord section (chord faces) and subsequently tearing of the chord side walls.
  With the T-type connection there occurs:
  —cracking at the corners of the bracing or of the chord side walls;

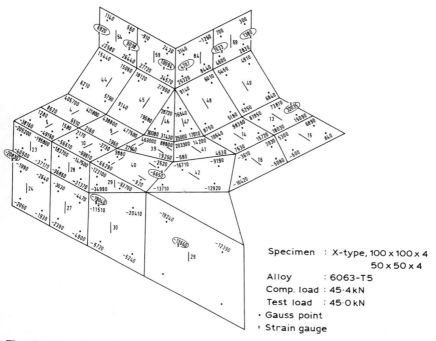

Fig. 24. Results of computer program and strain gauges in directions indicated; the strain values are given in μm/m.

—cracking at the toe of the weld of the bracing or cracking of the chord face.
- For both types of connection the highest strain values are measured where failure finally occurs.
- The strain values measured are much higher for the alloy 6063-T5 than for 7020-T6. With the latter alloy the highest strain values are restricted to the weld areas, whereas for the alloy 6063-T5 the highest strain values occur in the heat-affected zones, which causes better redistribution of stresses. This results in a larger effective width of the connection for the alloy 6063-T5, and explains the relatively high ultimate strength of the connection for the alloy 6063-T5 as compared with 7020-T6.
- For the design of such connections it is advisable to pay attention to failure not occurring in the welds with respect to the strength and deformation capacity of the connection. For the alloy 7020-T6 it

may be favourable to use a high-strength filler metal (for example 5280).

7 EVALUATION

All the results of the research programme described in this report, have been summarized and evaluated in Ref. 21. On the basis of these results, recommendations for the design of welded connections in aluminium alloy structures have been formulated which will be incorporated in Refs 1 and 3 as well as in 5.

The main differences in relation to the existing regulations are:

- Qualification of welders and welding procedures is necessary to apply the 'new' design rules.
- The strengths of the weld metal and heat-affected zone are dealt with separately.
- The design strength of the weld metal and heat-affected zone will be based on the respective ultimate strengths, see Tables 10 and 11.
- The design of butt welds is based on the design strength of the weld metal, whereas for fillet welds the design is based on the β-formula taking into account the design strength of the weld metal, the throat section and the direction of the different stresses in that section.
- To obtain sufficient deformation capacity of the welded connection it is advisable to aim for a higher strength of the welds compared with the strength of the heat-affected zone.

REFERENCES

1. ECCS, *European Recommendations for Aluminium Alloy Structures*, 1st edition, 1978.
2. DIN 4113, Aluminiumkonstruktionen unter vorwiegend ruhender Belastung, Teil 2: Geschweisste Konstruktionen, Berechnung und bauliche Durchbildung, 1980.
3. TGB-Aluminium, Technical principles for the design of building structures, aluminium structures. Netherlands Standard NEN 3854, 1983 (will be replaced by NEN 6710 to be published in 1991).
4. Soetens, F., Welded connections in aluminium alloy structures. Paper presented at the Second International Conference on Aluminium Weldments, Munich, 1982.

5. CP 118, *The Structural Use of Aluminium.* British Standard Code of Practice, British Standards Institution, London, 1969. (Will be replaced by BS 8118, to be published in 1992).
6. DTU 32/2, Règles de conception et de calcul des charpentes en alliages d'aluminium. Travaux de bâtiment, 1976.
7. Werner, G., Tests on welded connections of aluminium alloys. Final report, Doc. IIW-XV-328-72, 1972.
8. Werner, G., Aluminium-Schweissverbindungen. Untersuchungen an Schweissverbindungen von Aluminiumlegierungen des konstruktiven Ingenierbaues unter vorwiegend ruhender Beanspruchung, FMPA Baden-Württemberg, 1980, (Also Doc. IIW-495-81, 1981.)
9. Pirner, M., Properties of gas-shielded arc-welded joints in heat-treatable AlMgSi and AlZnMg alloys. Alusuisse paper for IIW colloquium in Oporto, 1981.
10. Fougeras, B., Charleux, J. & Brillant, M., Essais de soudage sur alliages corroyés, essais de rupture sur assemblages soudés en bout a bout, cisaillement et arrachement. Report of Aluminium Pechiney No. 860, 1968.
11. Soetens, F., Welded connections in aluminium alloy structures. TNO-IBBC Report No. BI-81-27, 1981.
12. Soetens, F., Welded connections in aluminium alloy structures, heat-affected zone effects. TNO-IBBC Report No. BI-84-17 by order of ECCS-T2, 1984.
13. Soetens, F., Welded connections in aluminium alloy structures, mechanical properties. TNO-IBBC Report No. BI-83-4 and BI-83-24, 1983 (in Dutch).
14. Fisher, R. A., *The Design of Experiments,* 6th edn. Oliver & Boyd, Edinburgh, 1951.
15. Gales, A., Inspection of welding fillet welds for the project on welded connections in aluminium alloy structures. TNO-MI Report No. 02956 M/GAA/NDR, April 1983 (in Dutch).
16. Soetens, F., Welded connections in aluminium alloy structures, progress report on fillet welds. TNO-IBBC Report No. BI-83-35, 1983 (in Dutch).
17. van Douwen, A. A. & Witteveen, J., Proposal to modify the ISO formula for the design of welded joints into a formula similar to the Huber-Hencky yield criterion (in Dutch). *Lastechniek,* **6** (1966).
18. Gales, A., Welded connections in aluminium alloy structures, assistance welding test series 8 and 9. TNO-MI Report, 1983 (in Dutch).
19. *Finite element system DIANA, Part III—Users' Guide.* TNO-TBBC, 1981.
20. Soetens, F. & de Coo, P. J. A., Welded connections in aluminium alloy structures, welded connections in hollow sections. TNO-IBBC Report No. BI-84-45, 1984 (in Dutch).
21. Soetens, F., Welded connections in aluminium alloy structures, evaluation and recommendations. TNO-IBBC Report No. BI-84-74, 1984 (in Dutch).

6

Joints with Mechanical Fasteners

M. S. G. CULLIMORE

Formerly University of Bristol, UK

ABSTRACT

The design of bolted and riveted joints made with single or with groups of fasteners to resist static loads is discussed for axially and eccentrically applied cases. The behaviour of the fastener and of the plies is dealt with for both serviceability and ultimate loading conditions. The limited available range of fatigue data is reviewed.

The stress distribution in pinned joints and lugs is described and a simple static design method is presented. The fatigue resistance of lugs to both constant and variable amplitude loadings is considered in fracture mechanics terms.

The mechanics of load transfer by interface friction in HSFG bolted joints and the factors affecting their design are discussed, together with the development of a high performance aluminium/steel joint. Particular attention is given to the effects of fretting on the fatigue resistance of HSFG bolted joints.

NOTATION

Symbols which are used once and defined in the text are not listed.

a	Crack length
$A_f(A_p)$	Cross-sectional area of fastener (ply)
c	Value of e for full bearing resistance of ply
$d_f(d)$	Diameter of fastener (hole)

e	End distance of fastener
f_n	Nut flexibility
$f_p(f_z)$	Axial (through thickness) flexibility of ply
$f_s(f_t)$	Flexibility of fastener in transverse shear (axial tension)
g	Distance of line of action of applied load from centroid (centre of rotation) of group of fasteners
K	Stress intensity factor
K_c	Critical value of K
K_t	Stress concentration factor
ΔK	Range of K in fatigue loading
ΔK_{th}	Threshold value of ΔK
L, b	Joint dimensions
m, C	Material parameters in crack growth law
n, n_p, n_l	Numbers of fasteners in a group
N	Number of fatigue loading cycles
$p(l)$	Longitudinal (lateral) spacing of fasteners
P_t, P_b, P_s	Resistance of ply to axial tension, bearing, shear (tear-out)
r_i	Distance of ith fastener from centroid (centre of rotation) of group of fasteners
R	Fatigue stress ratio
$R_t(R_s)$	Resistance of fastener to axial tension (shear); suffices: L = limiting, u = ultimate
s	Distance of centre of rotation from centroid of fastener group
S	External load on a joint
S_s	Applied load to cause slipping of HSFG bolted joint
t	Ply thickness
T	Bolt pre-tension
Y	Traction—tangential force between plies
z	Interface pressure in HSFG bolted joint
Z	Normal force on joint interface element
α	Geometrical factor in expression for K
β, ψ	Angular measurements
$\delta(\phi)$	Relative axial (angular) displacement between plies
θ	Coefficient of thermal expansion
Θ	Tension/temperature coefficient
λ	Traction coefficient

μ	Slip factor
$\sigma(\tau)$	Direct (shear) stress
σ_a	Amplitude of fatigue stress fluctuation
$\sigma_{av}(\sigma_g)$	Average (greatest) stress on cross-section
$\sigma_e(\tau_e)$	Stress at the elastic limit
σ_L	Limiting (design) stress
σ_m	Mean stress in fatigue stress cycle
$\sigma_{max}(\sigma_{min})$	Maximum (minimum) stress in fatigue loading cycle
σ_r	Fatigue stress range
σ_u	Ultimate stress
σ_y	Yield stress
$\sigma_{0.2}$	0·2% proof stress

1 INTRODUCTION

Well designed joints are essential to ensure the satisfactory performance of any structure. In aluminium frameworks with riveted or bolted gusset plates it has been estimated[1] that the weight of the joints is about 10% of the weight of the structure; in cost terms, the ratio is probably larger. A significant weight advantage results from the use of welding which reduces this ratio to about 4%. Welding may also be preferred for general engineering purposes because it simplifies fabrication and assembly, which reduces cost. However, where site assembly is required, or when the structure is subjected to fatigue loading, joints with mechanical fasteners—bolts or rivets—may be necessary. Furthermore, such joints provide useful system damping which is virtually absent in continuous welded structures.

The behaviour of structural joints has not attracted the research interest that its importance in structural performance would seem to merit. The bulk of research effort for the construction industry has been concerned with steel joints, comparatively little having been directed to problems peculiar to aluminium joints. However, the change from permissible stress to limit state design methods during the last decade, and the consequent need to revise codes of practice, has revived research interest in joint behaviour. Although limit state design, by definition, requires plastic (non-linear) analysis of structural behaviour, elastic analysis is still needed for the calculation of deformations in the serviceability limit state and, as will be discussed later, in fatigue life estimations.

The factors affecting the design of joints with mechanical fasteners are reviewed in this chapter, dealing first with simple single-fastener joints subjected to static in-plane (shear) loading, and then extending the discussion to joints with groups of fasteners subjected to translation or to rotation between the plates. A brief discussion is given of the case of loading normal to the plane of the joint which induces axial tension in the fasteners.

Pinned joints with single pins (lugs) are widely used, particularly in demountable structures and in structures with cable elements, and problems relating to their design for resistance to static and fatigue loading are examined.

The only major innovation in fastener technology since the development of welding has been the reliable use of friction in the transfer of load between joint interfaces. The existence of this effect, produced by the clamping action of hot driven rivets, has long been recognised, but its quantification for use in design, made practicable by accurately pre-tensioned high-strength bolts, has been developed only in the last 40 or so years.

The factors affecting the static resistance of friction grip bolted joints—interface friction, the control of bolt pre-tension, and joint geometry—are discussed in detail. The mechanism of interface fretting which has a major influence in determining fatigue life is discussed, together with geometrical and other factors affecting fatigue endurance.

It will be assumed throughout that the material is an aluminium alloy, except where it is stated to be otherwise.

Data are given in the units used in the original source material; except for these cases, SI units are used.

2 SINGLE FASTENER JOINTS WITH IN-PLANE LOADING

In simple design calculations the ultimate applied load is assumed to be shared equally between all the fasteners in the joint. The problem is therefore reduced to the design of an individual fastener and the adjacent joined material. It will, however, be shown later that these simple assumptions are inadequate in describing the behaviour of groups of fasteners.

The design parameters for single fastener joints are the resistance of the fastener to axial tension, shear and bearing, or to combinations of these actions and the resistance of the joined material to tension, tear-out (shear) and bearing.

2.1 Fastener resistance

2.1.1 Design stresses

The limiting axial tension stress (σ_L) used for design in aluminium alloy and stainless steel has a value between the 0·2% proof stress ($\sigma_{0.2}$) and the ultimate stress (σ_u) depending on the form of the stress–strain curve for the particular material. For mild steel and low alloy steels σ_L is the yield stress.

The elastic limit in shear (τ_e), determined by the Mises–Hencky criterion, is

$$\tau_e = \sigma_e/\sqrt{3} = 0.577\sigma_e$$

where σ_e is the elastic limit in uni-axial tension. The limiting shear stress for design (τ_L) is commonly taken as $0.6\sigma_L$ for aluminium and $0.7\sigma_L$ for steel.

2.1.2 Axial tension

The limiting resistance of a fastener to axial tension (R_{Lt}) is

$$R_{Lt} = \sigma_L A_f \tag{1}$$

where A_f is the cross-sectional area of the fastener effective in resisting the applied tension. The UK practice with steel bolts is to make A_f the tensile stress area instead of the thread core area. This larger area, calculated by using the mean of the effective and minor thread diameters, takes advantage of the enhancement in ultimate tensile strength resulting from the constraint effect of the tri-axial stresses in the thread roots.

2.1.3 Transverse shear

The limiting resistance of a fastener to transverse shear (R_{Ls}) is

$$R_{Ls} = \tau_L A_f \tag{2}$$

In this case A_f is the area of the cross-section of the fastener in the shear plane.

2.1.4 Axial tension and shear

For combined axial tension stress (σ) and shear stress (τ), the elastic limit is reached when

$$\sigma^2 + 3\tau^2 = \sigma_e^2$$

where σ_e is the elastic limit in uni-axial tension. Putting $\tau_e = \sigma_e/\sqrt{3}$ gives

$$\left(\frac{\sigma}{\sigma_e}\right)^2 + \left(\frac{\tau}{\tau_e}\right)^2 = 1 \tag{3}$$

which leads to the convenient interaction rule for design:

$$\left(\frac{\sigma}{\sigma_L}\right)^2 + \left(\frac{\tau}{\tau_L}\right)^2 \leq 1 \tag{4}$$

where $\tau_l = 0.6\sigma_L$ for aluminium and $0.7\sigma_L$ for steel. Fisher & Struik,[2] quoting the results of experiments with high strength steel bolts, give an equation for the interaction curve

$$\left(\frac{x}{0.62}\right)^2 + y^2 = 1 \tag{5}$$

where x = shear stress/tensile strength, and y = tensile stress/tensile strength.

2.1.5 Bearing of fasteners

A complex stress situation exists at the interface between the fastener and the surface of the plate (ply) through which it passes. The actual shear, bending, and radial compressive stresses, which vary over the area of contact, are usually represented by an average 'bearing stress' over the projected area = ply thickness (t) × fastener diameter (d_f). A conservative value of $2\sigma_L$ is commonly taken for the limiting bearing stress. On this basis the bearing resistance of a fastener in double shear will be greater than its shear resistance provided d_f is less than about $2t$; the precise value will depend on the hole clearance.

2.1.6 Resistance of the connected ply

The mode of failure of the connected ply can be either in tension across the minimum cross-section (A_p), by elongation of the hole (Fig. 1) by bearing, or by tear-out (shear) towards the end of the ply. The

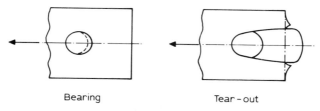

Fig. 1. Plate failure—single fastener.

resistances to these modes of failure are:

Tension resistance of the ply, $P_t = A_p \sigma_u$
Bearing resistance of the ply, $P_b = c_1 t d_f \sigma_u$ (6)
Tear-out resistance of the ply, $P_s = c_2 e t \sigma_u$

where c_1 and c_2 are coefficients relating the bearing and shear strengths of the material of the ply to its ultimate strength in tension (σ_u), which take account of the effects of the thinness (d_f/t) of the ply.

The minimum end distance (c) to develop the full bearing resistance of the ply, i.e. when $P_b = P_s$, is

$$c = \frac{c_1}{c_2} d_f \quad (7)$$

Marsh[3] states that, for aluminium plies of 'normal' thickness, ($d_f/t < 10$), a conservative minimum value of c is $2d_f$. For 'thinner' material ($d_f/t > 13$), $c = 1 \cdot 5 d_f$. For intermediate thicknesses, $c = (d_f/6)(22 - d_f/t)$.

2.1.7 Lap joints

Lap joints, in which the fasteners are in single shear, are found in truss gusset plates, in seams in plated structures, and in secondary members. They are simple to fabricate and erect but, because of the inherent eccentricity of the load (Fig. 2), are subjected to local

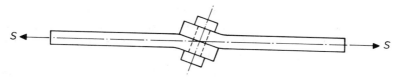

Fig. 2. Secondary bending in a lap joint under axial tensile load S.

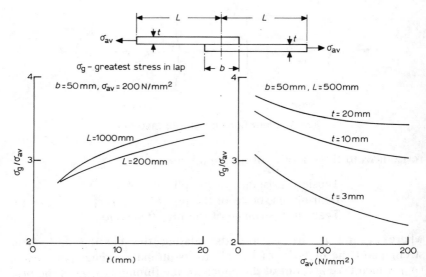

Fig. 3. Stresses in an aluminium alloy lap joint.

out-of-plane bending which also causes axial tension in the fastener. The greatest bending stresses occur near the ends of the lap, and are most pronounced with single fastener joints in short members.

A method of calculating the stresses in lap joints has been proposed by Cox,[4] in which the results are presented in terms of the non-dimensional parameters $(\sigma_{av}/E)(L/t)^2$ and (L/b), where b is the length of the lap over which firm contact is maintained (see Fig. 3). Provision is also made for the restraint to rotation of the centre of the lap by the elements of the structure to which the member is attached and, in the case of seams in cylinders, by curvature of the plates. The effect of such restraint in a particular case is to reduce the maximum stress in the joint (σ_g).

The parameters affecting the stress magnification factor—maximum stress $(\sigma_g) \div$ average tensile stress (σ_{av})—are interdependent, so trends are most easily discussed in terms of a particular configuration. Figure 3 shows that (σ_g/σ_{av}) increases as the plate thickness increases, and that a small reduction is caused as L decreases (Fig. 3(a)). This decrease is, however, restricted by the lower limit imposed on the length (L) by practical considerations. Reducing the axial stress increases the stress magnification (Fig. 3(b)).

In most practical cases these secondary stresses reduce the static ultimate strength by up to about only 10%, provided the integrity of the joint is maintained by using an adequately proportioned fastener providing some grip, such as a pre-tensioned bolt or squeeze driven rivet. If bending of the joint is adequately restrained, as is the case when two T-sections are joined back-to-back, secondary effects can be neglected in calculating the ultimate strength. In fatigue loading, however, the secondary bending stresses under service loading can significantly reduce the fatigue endurance and must be taken into account.

3 JOINTS WITH GROUPS OF FASTENERS

3.1 In-plane axial loading

The problem of the partition of load between a line of fasteners in a joint, discussed by Batho[5] in 1916, has subsequently been the subject of many analytical and experimental investigations. A study, specifically related to aluminium alloy joints, has been reported by Francis.[6] A basis for the design of riveted joints in aluminium alloy ships' plating, taking account of strength and watertightness, proposed by Flint,[7] was based on tests on large lap and cover plate joint specimens with cold driven N6 (5056A) aluminium rivets and hot driven mild steel rivets.

3.1.1 Elastic distribution of fastener resistance
The method of elastic analysis is essentially common to all the authors, and will be presented here, for simplicity, in terms of a double cover plate butt joint (Fig. 4(a)). The fasteners transfer the external load (S) from the centre to the outer plates. In a segment of the joint between two fasteners, shown schematically in Fig. 4(b), the forces in the two plates, and hence their longitudinal extensions, will generally be different. This difference in extensions will be accommodated by the shear deformations of the fasteners and local deformation of the adjacent plate material, δ. Generally δ is a function of R_s, the fastener shear resistance, so that

$$\delta = f(R_s)$$

A typical form of the relationship is shown in Fig. 5. For the early part, the relationship is approximately linear-elastic and the fastener

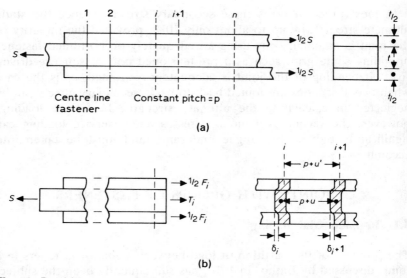

Fig. 4. Double cover plate butt joint. (a) Joint geometry. (b) Joint segment $i - (i+1)$.

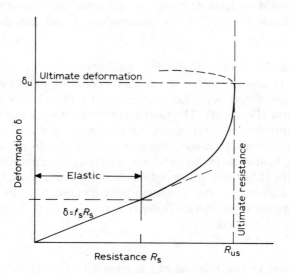

Fig. 5. Fastener resistance/deformation relationship.

flexibility (f_s), i.e. shear deformation per unit fastener resistance, is constant, so that

$$\delta = f_s R_s \qquad (8)$$

For equilibrium of external and internal forces at a section through the segment between the ith and $(i+1)$th fasteners

$$S = T_i + F_i$$

But

$$F_i = \sum_{i=1}^{i} R_{si}$$

where R_{si} is the resistance of the ith fastener, so

$$T_i = S - \sum_{i=1}^{i} R_{si} \qquad (9)$$

The extensions of the plates, in this segment are:

$$\text{centre plate, } u = T_i p / AE = f_{pi} T_i$$
$$\text{outer plate, } u' = F_i p / AE = f_{pi} / T_i \qquad (10)$$

where $f_{pi} = p/AE$ is effectively constant throughout the ith segment, and $T_i > F_i$ for this segment.

For compatibility of displacements

$$\delta_i + u'_i - u_i = \delta_{i+1}$$

Now putting $\delta_i = f(R_i)$ and $\delta_{i+1} = f(R_{i+1})$ and substituting for u_i and u'_i from eqn (10) in terms of R_i, etc., from eqn (9) gives

$$f[R_{s(i+1)}] = f(R_{si}) + f_{pi}\left[S - 2\sum_{i=1}^{i} R_{si}\right] \qquad (11)$$

For a joint with n fasteners there will be $(n-1)$ of these equations which, with the equation for overall equilibrium

$$S - \sum_{i=1}^{i=n} R_{si} = 0 \qquad (12)$$

can be solved to give the fastener resistances. For elastic conditions, eqn (8), $\delta = f_s R_s$. In the simple case where the plates are of constant width and thickness and all the fasteners are the same, f_s and f_p are constant throughout the joint, thus eqn (11) becomes

$$R_{i+1} = R_{si} - \frac{f_p}{f_s}\left|S - 2\sum_{i=1}^{i} R_{si}\right| \tag{13}$$

It can be seen from eqn (13) that up to the middle of the joint, where $T_i = F_i$, R_{si} is greater than $R_{s(i+1)}$. Thus the fastener resistance is greatest at the ends of the joint and decreases towards the middle. Francis[6] (loc. cit) showed that for $\frac{3}{4}$-in diameter N6 (5056A) aluminium rivets in H15W (2014A T4) aluminium plate in elastic conditions, the ratio of the greatest rivet load to the average load (S/n) with 3, 5 and 12 rivets was 1·17, 1·58 and 2·8 respectively. If the area of the outer plates is greater than that of the centre plate, the variation of fastener resistance is skewed, the first fastener exerting a greater resistance than the last. Reducing the Young's modulus of the plates will increase f_s (eqn (8)) and also increase the deformation of the plate by the fastener. The latter, however, is generally the lesser part of the fastener flexibility (f_s), the increase in which will be proportionally lower. The net effect is to increase the factor (f_p/f_s) in eqn (13) so that, for geometrically similar joints, using the same fasteners, the variation in the fastener resistances will tend to be more pronounced with aluminium than with steel joints. An increase in the pitch (p) will also increase f_s, with the same effect.

Although the distribution of fastener resistance in elastic conditions is not required for static limit state design, it will be shown later, particularly in friction grip bolted joints, that in fatigue loading the endurance indicated by the average fastener resistance may be reduced by the effects of the higher local stresses and displacements which occur at the end fasteners of long joints. A knowledge of the elastic distribution of fastener resistances may therefore be needed for design against fatigue.

3.1.2 *Fastener flexibility*
The elastic analysis described above presents no problems using a simple computer program to solve the resulting set of simultaneous equations, provided that the fastener flexibility (f_s) is known. The fastener flexibility is a function of the properties of the fastener and of the plates on which it bears, and will be different for each particular fastener/plate combination. It may be obtained by direct measurement but, because of the high cost, this is not generally feasible except for large-scale projects.

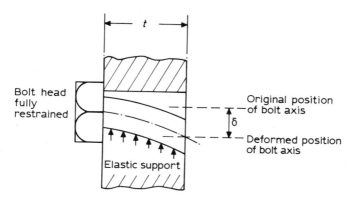

Fig. 6. Idealised deformation of bolt in a shear joint.

A method of estimating f_s for bolts, based on small deflection theory, resulted from an investigation at the British Aircraft Corporation for the UK Ministry of Defence.[8] The bolt is treated as a cylinder supported on an elastic foundation (Fig. 6), and the result is expressed in terms of the shear and bending of the bolt, the through-thickness flexibility (f_z) of the supporting plate, and the rotational restraint provided by the bolt head. A computer program is provided which calculates all the parameters to be varied and gives a result (f_z/f_s) in terms of (t/d_f). The curves presented use an empirical relation $f_z = 5\cdot 56 \div$ Young's modulus for the plate appropriate to titanium bolts in aluminium plates. This gives reasonable agreement with test results for steel bolts if $(t/d_f) > 1\cdot 75$. If better agreement is required, or for other values of (t/d_f), a simple test for measuring f_z is described.

The bolt flexibility is least when rotation of the bolt head is fully restrained—a condition which is approached when the bolt is fully pre-tensioned. Reducing the head restraint causes f_s to increase, the increase being largest at low values of (t/d_f).

3.1.3 Ultimate strength

Provided the full fastener resistance/deformation relationship has been obtained, either by experiment or empirically, the elastic analysis can be adapted for use in the much more complex non-linear computation required for an ultimate strength analysis. When the resistance of a fastener reaches its maximum value, R_{us} (Fig. 5), the fastener will continue to deform until the limiting deformation (δ_u) is reached, when rupture will occur. After the external applied load has exceeded the value to cause the most heavily loaded fastener to reach its maximum resistance (R_{us}), the distribution of fastener resistance tends to become less uneven. If the ductility of the fasteners was unrestricted, the applied load could be increased until all the remaining fasteners had reached their maximum resistances, giving an ultimate load of $S_u = nR_{us}$.

When the number of fasteners in the joint is small, and the joint is compact, the difference between the highest and the average fastener load is small, and the ductility of the fasteners is normally sufficient almost to equalise their resistances at the ultimate load. However, in long joints the limiting deformation of the end fasteners (δ_u) may be reached and failure may result before the middle fasteners have developed their maximum resistances. The average fastener resistance for the joint, $R_a = S_u/n$, is then less than R_{us}. Unintentional variations in pitch will result in some fasteners bearing on the plates before others. This will distort the initial load distribution, but should normally have little effect on the ultimate resistance.

A study of bolted steel joints by Fisher & Rumpf[9] showed that with up to four bolts ($n = 4$) the decrease in R_a was negligible; for $n = 11$, $R_a = 0.80R_u$, and for $n = 22$, $R_a = 0.66R_u$. The rate of decrease of fastener efficiency, R_a/R_u, decreased as n increased. The yield strength of the plate material used was 36 ksi and the UTS of the $\frac{7}{8}$-in diameter bolts was 120 ksi. The bolt pitch was 3·5 in.

Francis[10] concluded from an extensive investigation on aluminium riveted joints, with HP15W plate ($\sigma_{0.2} = 290 \text{ N/mm}^2$, $\sigma_u = 450 \text{ N/mm}^2$) and N6 rivets ($\sigma_{0.2} = 140 \text{ N/mm}^2$, $\sigma_u = 270 \text{ N/mm}^2$), that with well proportioned joints the loss of rivet efficiency was negligible when the number of rivets in a line did not exceed eight, and the rivet pitch was less than $6d_f$. Rivet efficiency improved with increasing the d_f/t ratio up to a value of 1·5.

3.1.4 *Resistance of the plies*

It has been shown by Marsh (loc. cit) that the true ultimate ply bearing capacity of a multi-fastener joint is a function of the fastener spacing, and can be overestimated by considering the resistance at an individual fastener only.

In calculating the resistance to tear-out (P_s) of a group of fasteners in a joint (Fig. 7) shearing along the lines AD and BC, tension across CD must be considered, together with tear-out of the last row of fasteners towards the unloaded end of the ply along AA' and BB'. If

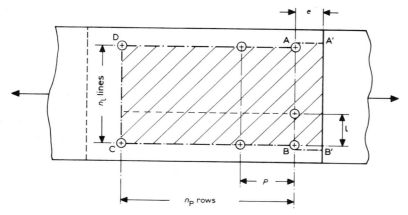

Fig. 7. Tear-out of a group of fasteners.

there are n_p rows, each with n_l fasteners all of diameter d_f, then

$$P_s = (n_l - 1)(l - d_f)t\sigma_u + 2(n_p - 1)(p - d_f)t\tau_u + et\sigma_u. \quad (14)$$

It is assumed that $\tau_u = 0.5\sigma_u$ and that $e = c$ (eqn (7)), the value required to develop full ply bearing resistance ($ct\sigma_u$) at each fastener. The ply bearing resistance for the whole joint is therefore

$$P_b = n_l n_p ct\sigma_u \quad (15)$$

Equating P_s from eqn (14) and P_b from eqn (15) gives the condition for equal resistance to tear-out and bearing

$$(n_l - 1)(l - d_f) + (n_p - 1)(p - d_f) + c = n_l n_p c \quad (16)$$

Equation (16) is satisfied if $l = \frac{1}{2}(n_p + 1)c + d_f$ and $p = \frac{1}{2}(n_l + 1)c + d_f$.

In the case of an aluminium alloy joint having four rows of five fasteners each at the minimum spacing of $2.5d_f$ and $c = 2d_f$, eqn (14) gives a tear-out strength of $12.5td_f\sigma_u$, whereas the aggregate of the individual bearing resistances ($n_l n_p ct\sigma_u$) is $40td_f\sigma_u$. To obtain full bearing resistance, eqn (16) indicates that a lateral spacing of $6d_f$ and a longitudinal spacing of $7d_f$ is required.

Tests on joints containing up to six fasteners gave close agreement with predictions of strength by this method, confirming that the aggregate of the bearing resistances at the individual fasteners can overestimate the resistance to tear-out of joints with compact fastener groups.

3.1.5 Factors affecting design

For optimum ultimate strength performance, joints should be proportioned to give the most uniform distribution of fastener resistance. This is likely to be achieved by keeping the joint short, using the smallest number of rows of fasteners and having the joined parts of equal cross-section. High fastener ductility increases the uniformity of fastener resistance; a factor which should be noted when using high tensile steel bolts in aluminium alloy plates.

In order to maintain full bearing resistance of the joint before tear-out, for each additional row of bolts, $n_p \to (n_p + 1)$, the lateral spacing (l) must be increased by $c/2$.

The variety of both geometrical and material strength parameters involved in the behaviour of long joints makes it impossible to cover effectively all joint configurations and alloys with a single simple formula suitable for use in a code of practice. As a guide, limiting the number of rows of fasteners to five before deductions for the long joint effect need to be considered seems to be safe, if somewhat conservative. This rule, combined with the restriction of the overall length of the joint to $15d_f$ (pitch $3 \cdot 75d_f$ with five fasteners), could be expected to restrict the loss of fastener efficiency in bolted and riveted joints to about 1%.

3.2 In-plane eccentric loading

When the resultant of the external forces on a joint does not pass through the centroid of the fastener group the joint is subjected to axial force and moment, so that the fastener resisting forces are no longer all in the direction of the resultant external force.

3.2.1 Elastic analysis

The resultant force, S, on a group of n similar fasteners (Fig. 8) can be replaced by an equal and opposite parallel force through the centroid, G, of the group, and a moment, Sg, about G. The simple elastic analysis is based on the assumption that the joined plates are rigid and undergo a relative rotation, ϕ, about G that is the same for all fasteners. The resistance (U_i) to the moment provided by the fastener at I is assumed to be proportional to the relative displacement of the

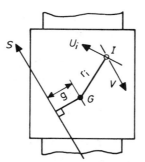

Fig. 8. Eccentric loading—elastic analysis. G is centroid of group of n fasteners.

plates at its centre, $r_i\phi$, where $r_i = $ GI, and to be in a direction perpendicular to GI, so that

$$U_i = kr_i\phi \qquad (17)$$

where k is a stiffness coefficient which is the same for all fasteners.

The resultant resisting moment is

$$\sum_{i=1}^{i=n} U_i r_i = Sg,$$

Hence, substituting for U_i from eqn (17)

$$k\phi = Sg \Big/ \sum_{i=1}^{i=n} r_i^2$$

so that

$$U_i = k\phi r_i = r_i Sg \Big/ \sum_{i=1}^{i=n} r_i^2 \qquad (18)$$

The resistance of each fastener to the force through the centroid is $V = S/n$, parallel to S. These two components U_i and V are then added vectorially to give the resultant resistance of the fastener R_{si}. The most heavily loaded fastener will be furthest from the centroid of the group on the side towards the external load.

The method can be expected to give satisfactory results for well-fitting fasteners in stiff plates at loads below the elastic limit and, in these conditions, to be useful in some fatigue loading calculations.

3.2.2 Ultimate load analysis

Predictions of the ultimate load using the elastic method are known to be conservative, giving values for steel plates as much as 20–35% larger than values obtained from tests. Rational methods have been developed[11,12] which consider the actual load–displacement relation for the single fastener. The joined plates are assumed to undergo relative rotation about an instantaneous centre, C, whose position, for a particular load, is initially unknown (Fig. 9). The plates are assumed to remain rigid so that the shear deflection of a fastener is proportional to its distance from the centre of rotation. The ultimate load is that causing the ultimate deflection (δ_u) in the fastener furthest from the instantaneous centre.

The distance, r_i, of the fastener at I from the centre of rotation, C, whose co-ordinates are taken to be x_c, y_c, is

$$r_i = \{(x_c - x_i)^2 + (y_c - y_i)^2\}^{1/2}$$

The x and y components of the shear resistance of the fastener, R_{si}, which is obtained from its non-linear load–displacement relation (R_s/δ) (see Fig. 5), are:

$$H_i = R_{si}(y_c - y_i)/r_i, \qquad V_i = R_{si}(x_c - x_i)/r_i$$

For equilibrium of internal and external forces,

$$H = \sum H_i \quad \text{and} \quad V = \sum V_i$$

so that

$$H/V = \sum H_i / \sum V_i = \cot \beta \text{ (the known direction of S)} \tag{19}$$

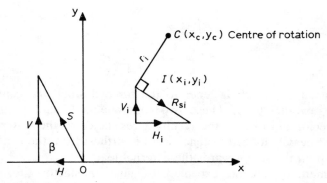

Fig. 9. Eccentric loading—ultimate load analysis.

For overall equilibrium of moments about C

$$x_c V + y_c H - \sum r_i R_{si} = 0$$

or (20)

$$x_c \sum V_i + y_c \sum H_i - \sum r_i R_{si} = 0$$

The solution is obtained by iteration to arrive at values of x_c, y_c, the co-ordinates of the centre of rotation, which simultaneously satisfy eqns (19) and (20). This method is similar in principle to that for the analysis of long joints and, likewise, requires a knowledge of the fastener deformation/resistance relationship, which can be found experimentally. This is usually done by a shear box type test, in which the fastener is loaded transversely in double shear. The test loading conditions differ from those experienced by the fasteners in a joint where the direction of the load on the fastener changes as the load increases. This is caused by movement of the position of the instantaneous centre which results in some rotation of the plates around the fastener and causes its resistance to be less than that found in the shear test.

In an extensive study of bolted steel joints, Crawford & Kulak[11] (loc. cit.) devised an expression for the fastener deformation/resistance curve with empirical coefficients to fit experimental data for $\tfrac{3}{4}$-in diameter A325 bolts. Design data in the form of tables of values of a coefficient, Q, were derived, from which the ultimate value of the eccentric load (S_u) for one and two rows of bolts can be found from the expression

$$S_u = Q A_f \tau_u \qquad (21)$$

where τ_u is the shear strength of the bolt material. These data may be used for any carbon or low alloy steel plate as variations of plate strength only affect the amount of total deformation to which the ultimate strength is not sensitive. It was also found that the result was independent of bolt diameter within the common range of structural sizes. The design data are not applicable to the higher strength grade bolts which have different deformation characteristics.

The plate deformation component of the total fastener deformation will be much greater in aluminium than in steel plates. New data will therefore have to be derived for this method to be used for the design of eccentrically loaded aluminium joints.

A simplified method for the design of eccentrically loaded bolted aluminium joints has been proposed by Marsh[13] which is claimed to give results within some 10% of the successive iteration method, and is used in a proposed ISO standard for the design of aluminium alloy structures. It is based on the elastic method for finding the most severely loaded fastener.

When a force, S, acts at a distance g from the centroid of a bolt group, the distance s of the instantaneous centre of rotation from the centroid is $s = r_p^2/g$, where r_p is the polar radius of gyration of the bolt group relative to the centroid. The resisting force of the fastener at a distance r_i from the centre of rotation is

$$R_{si} = S(s+g)r_i \bigg/ \sum_{i=1}^{i=n} r_i^2 \qquad (22)$$

When all the fasteners in the group have reached their ultimate resistance, R_{us}, the ultimate value of the applied load is

$$S_u = \left(\frac{R_{us}}{g+s}\right) \sum_{i=1}^{i=n} r_i \qquad (23)$$

Compared with Crawford and Kulak's results (loc. cit.), this method is unconservative; but Marsh states that using the mean value of the ultimate bolt strength minus two standard deviations brings the results into agreement. The validity of this method will also depend upon there being sufficient ductility in the material for simultaneous development of the ultimate conditions at each fastener. These conditions are most likely to be obtained in compact bolt groups.

3.3 Fatigue loading

Joints with mechanical fasteners are free of some of the consequences of fabrication which affect the fatigue strength of welded joints, notably high tensile residual stresses and, in the case of heat-treatable alloys, reduction of strength in the heat-affected zone of the weld. Residual stresses may, however, be induced in the fabrication of bolted and riveted joints by forcing together badly-fitting or distorted components. These, unlike welding residual stresses, are not generally quantifiable.

3.3.1 Bolted joints without pre-tension
Joints in which the fasteners act as pins, i.e. where there is no clamping action, fail in fatigue by cracking from the inside surfaces of

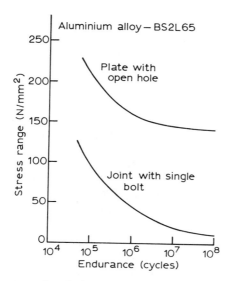

Fig. 10. Fatigue strength of single bolt butt joint compared with plain plate with hole.

the bolt holes. This is caused by the high local stress and fretting at the contact with the bolt shank there, and reduces the fatigue endurance below that of a plain plate with a similar hole. This has been demonstrated by Perrett[14] in tests with single bolt double cover plate butt joints in aluminium alloy BS2L 65 ($\sigma_{0.2} = 440\,\text{N/mm}^2$, $\sigma_u = 483\,\text{N/mm}^2$). Figure 10 shows that the fatigue strength of the joint is very much less than that of the plain plate with a hole, despite the fact that in the latter case the mean stress is higher (138 N/mm² cf. 110 N/mm²). Another significant feature, typical when fretting is present, is that even at $N = 10^8$ the σ_r–N curve for the joint has not yet reached an endurance limit.

Heywood[15] concluded from an analysis of 230 fatigue tests on typical bolted aluminium aircraft joints that the standard of design was the predominating influence on fatigue strength. About three times the fatigue stress range could be expected for a well designed joint, compared with one for which no special attention had been paid to fatigue. The influences of mean stress and alloy strength were found to be comparatively small.

3.3.2 Riveted joints

The process of driving a rivet will cause distortion of the plate and leave beneficial residual compressive stresses around the rivet hole. Aluminium rivets are normally driven cold, and so will exert less clamping action between the joint plies than hot driven steel ones. Little is known about the effects of bad workmanship on the fatigue strength of riveted joints. It will, however, be more apparent to an inspector than it is with welding, so that poorly-made riveted joints are likely to be rejected.

A series of fatigue tests were carried out by McLester[16] using specimens having a single row of two $\frac{5}{8}$-in diameter cold driven aluminium rivets in $\frac{1}{4}$-in thick plate. The alloy H30WP (6083 T6) was chosen as being likely to give results representative of the alloys normally used for riveting.

The fatigue strength of the joints was found to be significantly affected by mean stress, being lowered as the mean stress increased (Fig. 11) and, as would be expected because of the secondary bending stresses, the fatigue strength of the lap joints was less than that of the butt joints.

It was found that the principal cause of failure was fretting between the faying surfaces. For this to have been so, there must have been a

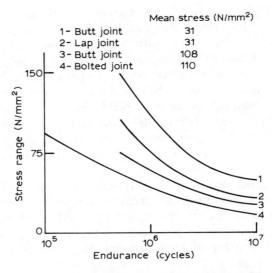

Fig. 11. Fatigue strength of riveted aluminium alloy joints.

degree of clamping between the plates and, consequently, a transfer of load by friction between them. This would tend to enhance the fatigue strength. The σ_r–N curve for mean stress 108 N/mm² lies above that for the bolted joint (Fig. 10) at a similar mean stress (110 N/mm²), which tends to support this argument. Riveted lap joints are normally used in lightweight structures of thin sheet construction. ESDU Data Item 79031[17] gives the results of over 1050 constant amplitude fatigue tests on lap joints in a variety of aluminium alloys using snap head and countersunk rivets. In most cases the rivets were of the conventional solid type.

Rivet failures rarely occurred at endurances over $N = 10^5$. Sheet failures were from cracks propagating parallel to the joint line either from the end of a transverse diameter or from the end of a longitudinal diameter of a hole, or from the outer edge of the rivet head. In one set of tests, the use of a synthetic rubber adhesive in the lap was found to increase the endurance.

For two and three rows of rivets the optimum geometrical design parameters were found to be $d_f/t = 3 \cdot 0$, $l/p = 0 \cdot 8$ and $l/d_f = 3$ to 4.

3.3.3 Spot welded joints

Spot welded joints, commonly used in light construction, behave in a generally similar way to riveted joints. Failure under static loading is normally by shearing of the welds or by the weld spots tearing out of the sheet; fatigue failure is usually in the sheet. Typical values of fatigue strength for both steel and aluminium, reported by Welter,[18] were between 10% and 15% of the static strength, increasing with size of spot, number of spots in a row and number of rows. Butt joints are to be preferred to lap joints, as bending stresses in lap joints have an adverse effect on their fatigue strength.

4 PINNED JOINTS

This type of joint, where the two elements are connected with a single pin, is widely used in both civil and aerospace structures. Pinned joints occur where some articulation is needed, such as at the terminations of cable elements in tension structures, and in modular structures, such as military bridges, where it is required to dismantle and re-erect the structure.

4.1 Static loading

4.1.1 Stress distribution

In the simple square-ended lug in Fig. 12, the tensile load is transferred from the pin by bearing. The state of stress in the pin is primarily shear with some bending. The amount of bending depends on the pin clearance, and the lateral clearance between the lug and the fork end or clevis forming the other half of the joint.

The distribution of stress in the lug is complex, and much of the earlier data which form the bases of elastic design rules are derived from classical analysis and photo-elastic model experiments.[19] This technique has now largely been replaced by finite element analysis. The maximum tension stress in a lug with a close-fitting pin occurs at the ends of the transverse diameter through the centre of the hole, and is significantly larger than that in a plane plate with the same size hole carrying the same load. As the pin clearance increases, the point of maximum stress ('A' in Fig. 12) moves towards the head of the lug; for

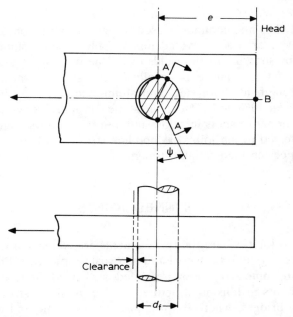

Fig. 12. Square-ended lug. •, Points of maximum stress.

normal clearance fits (<0·1%), ψ may be in the range 15–30°. Compression over the pin bearing area causes circumferential tension in the outside of the head of the lug. The maximum values of these stresses (at 'B' in Fig. 12) are less than the maximum stresses (A) at the hole boundary, except for small values of e/d_f. These features are illustrated in Fig. 13, which gives the results of the finite element analysis of a square-ended lug using the ANSYS program.[20]

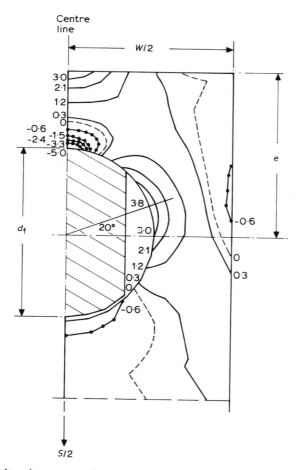

Fig. 13. Finite element analysis of circumferential stress distribution in a square-ended lug. Contours of local stress ÷ $S/(W-d)t$. $e/W = 0.50$, $d_f/W = 0.52$, (side relieved) pin clearance = 2·47%.

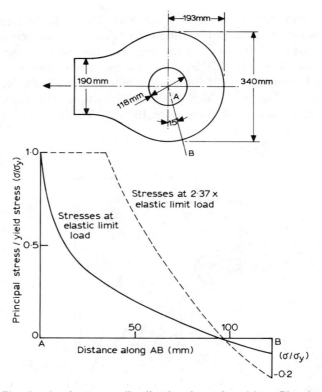

Fig. 14. Elastic-plastic stress distribution in waisted lug. Pin clearance = 0·65%; A, point of maximum stress.

Figure 14 shows the results of a finite element elastic-plastic analysis of a waisted lug using simple plastic elements.[21]

In elastic conditions the stress is greatest at the edge of the hole on a radius inclined at about 15° to the transverse diameter, and reduces very rapidly towards the outside of the lug where there is a small compression stress: this was also observed in the square-ended lug (Fig. 13). The stress distribution shown for a load of 2·37 times the elastic limit load indicates that slightly under 30% only of the transverse cross-section of the lug has reached yield.

4.1.2 Simple design

Failure of a lug may be by tension across the minimum section through the hole, by bursting of the head due to circumferential tension, by shear out of the pin, or by bearing. The latter, which elongates the hole, might be regarded as a serviceability limit.

The use of a finite element analysis for routine ultimate limit state design is clearly impracticable, except in special cases. A method recently proposed by Webber[22] gives the tension load carrying capacity based on net cross-sectional area, material properties and a geometry-dependent efficiency factor. Shear, bearing and bursting are covered by a single calculation based on the projected bearing area which includes a bearing efficiency factor which varies with the end distance and thickness ratios, e/d_f and t/d_f.

4.2 Fatigue loading

Lugs fail in fatigue from cracks initiating inside the pin hole due to the effects of the increased stress there aggravated by fretting between the pin and the inside surface of the hole. Because of this, the fatigue resistance of lugs is less than that of notched bars having the same stress concentration factor. The effect increases with higher endurances, as illustrated in Fig. 15.

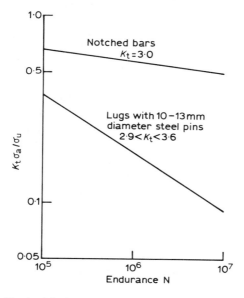

Fig. 15. Typical fatigue strength of Al–Zn–Mg alloy lugs.

Fatigue resistance is improved by moving the fretting area away from the points of maximum stress. This may be done by increasing the clearance of the pin in the hole, or by machining flat areas on the ends of its transverse diameter. Initial tensile overloading of the lug assembly, giving residual compression stress around the hole, increases the fatigue resistance. A similar effect may be achieved by initially expanding the hole with a mandrel or by forcing a ball through it ('ballising'). Steel pins used in aluminium alloy lugs are commonly cadmium plated to prevent bi-metallic corrosion between the steel and the aluminium, which would aggravate fretting. A large variety of anti-fretting compounds is commercially available.

4.2.1 Fatigue resistance of lugs under constant amplitude loading

A very large amount of research has been carried out on the fatigue of lugs, mainly in the aerospace industry, reflecting the importance attached to the design of this component. The results of some 1900 constant amplitude fatigue tests to failure on lugs of various geometries made of Al–Cu–Mg and Al–Zn–Mg alloys have been analysed by ESDU.[23] The lugs were loaded axially through nominally push-fit steel pins of diameters ranging from 5 mm to 60 mm having clearances ranging up to 1·2%. Some specimens with interference fit pins were included. No endurance improving techniques were used.

σ_a–N data for $10^4 < N < 10^8$ are presented in terms of the non-dimensional stress parameters $K_t\sigma_a/\sigma_u$ and $K_t\sigma_m/\sigma_u$, where the fatigue stress amplitude (σ_a) and the mean stress (σ_m) are calculated on the net area of the transverse cross-section through the centre of the hole. The stress concentration factor (K_t), which takes account of pin bending for the thicker lugs ($t/d_f > 0.5$), varied from 2·4 to 4·4. A decrease of mean stress tended to increase endurance, the effect being less pronounced at the higher mean stress levels. There was no conspicuous difference in the fatigue strengths of the Al–Cu–Mg alloy and the Al–Zn–Mg alloy specimens.

In many cases the local maximum stress in the lug at the upper load in the fatigue cycle exceeded the proof stress of the material. The endurance of these yielded specimens was generally found to be less than those that remained elastic. The limited data available for yielded Al–Zn–Mg alloy lugs suggested that their fatigue strength was much the same as that of comparable yielded specimens in Al–Cu–Mg alloy. A definite size effect is apparent in the results. For a given value of $K_t\sigma_a/\sigma_u$, the smaller pins have the superior endurance.

4.2.2 Safe life fatigue design of lugs

Fatigue design using σ_a–N data obtained from constant amplitude fatigue tests to failure employs partial safety factors on either stress or life, or possibly on both, to obtain a 'safe-life' result. In the case of lugs, which rely on a single element to transmit the load, fatigue failure could have unacceptable consequences, and so designers have tended to choose relatively larger safety factors than would be used for other elements.

4.2.3 Fatigue crack propagation in lugs: damage-tolerant design

A crack in a lug may be initiated by fatigue or start from an existing defect. Fluctuating stress will cause the crack to propagate until the remaining ligament cannot sustain the maximum service load, or until the crack reaches a critical length at which fast fracture occurs. This form of fracture is dangerous, as it does not permit the internal redistribution of load associated with ductile fracture. The determination of the critical crack length and the time taken for a crack to grow to this length are important parameters determining the safety of a structure. If the component is inspectable and a damage-tolerant design adopted, knowledge of the crack growth is necessary to fix the inspection intervals.

Fatigue crack growth data are presented in terms of the rate of growth per cycle (da/dN) and the range (ΔK) of the stress intensity factor (K). The stress intensity factor for a particular crack, of length a, is

$$K = \alpha \sigma \sqrt{\pi a} \qquad (24)$$

where σ is the average stress remote from the crack, α is a function of the crack length and the geometrical configuration of the component. For a crack in the centre of an infinite flat plate, $\alpha = 1$. Expressions for calculating stress intensity factors are available in handbooks, such as that by Rooke & Cartwright,[24] and in papers by them and by many other authors.

The form of a typical crack growth curve obtained from a standard test is shown in Fig. 16. Part 2–3 of the curve represents the region of static mode behaviour resulting in final failure at the critical stress intensity factor (K_c). Part 0–1 describes the region of non-continuum behaviour in which the rate of growth tends to zero at the threshold value (ΔK_{th}) of the stress intensity factor range. This value, below

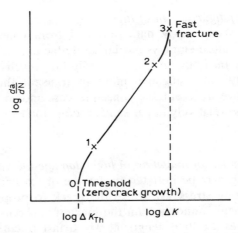

Fig. 16. Form of fatigue crack growth rate relation.

which crack propagation ceases, is lowered by increasing (R), the stress ratio $R = \sigma_{min}/\sigma_{max}$, and by corrosive environments.

Empirical and semi-empirical relations, fitted to experimental data, have been devised to represent crack growth. The best known of these is the Paris Law:

$$(da/dN) = C(\Delta K)^m \tag{25}$$

in which C and m are empirical material parameters. This expression fits the approximately linear intermediate part of the curve (1–2); the Forman Law:

$$(da/dN) = C(\Delta K)^m / [(1 - R)K_c - \Delta K] \tag{26}$$

describes part 1–2 and the upper part (2–3) also. An expression using an inverse hyperbolic function of ΔK with additional material parameters, devised more recently by Jaske et al.,[25] represents the whole curve.

The analytical prediction of crack growth in a component is possible for constant amplitude loading in the linear (1–2) region in cases for which the geometrical factor α is a simple function of crack length (a). For more complex crack growth laws and geometrical factor functions, a numerical integration approach must be used.

For actual structures it may be difficult to estimate ΔK accurately because of uncertainties about the shape of the crack and about other factors such as environment and load shedding in redundant structures. With these reservations, the procedures can be regarded as giving a useful first estimate of crack growth. The simple summation methods, if used for variable amplitude loading, will generally give conservative predictions because the effects of crack growth retardation are not taken into account. The effect of introducing occasional single high positive stresses in a constant amplitude loading sequence is to enlarge the plastic zone ahead of the crack tip, causing high compressive residual stresses which retard the extension of the crack under the subsequent lower stress cycles. This causes the crack growth curve to be discontinuous, as illustrated diagrammatically in Fig. 17. If, however, the peaks are frequent, so that their effects are superimposed on each other, the overall crack growth is slowed and the curve becomes effectively continuous. Descriptions of the methods for predicting crack growth under practical variable amplitude loading spectra will be found in most standard texts on fracture mechanics.

4.2.4 Experimental study of fatigue crack propagation in lugs under constant and variable amplitude loading

A report of an investigation by Buch & Berkovits[26] of crack propagation in aluminium alloy lugs describes tests on lugs having 10 mm and 30 mm diameter clearance fit pins, and contains a useful review of other work in this field.

Crack propagation experiments were carried out under constant

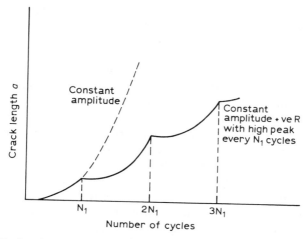

Fig. 17. Crack growth under simple variable amplitude loading.

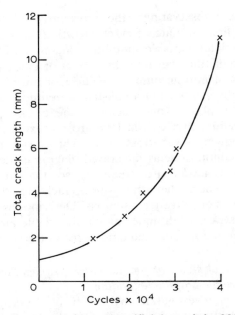

Fig. 18. Measured growth of 1 mm artificial crack in 2024-T3 lug under constant amplitude loading. Stress amplitude 31·5 N/mm^2; R = 0; pin diameter 30 mm.

amplitude and under variable amplitude loading using a randomised loading program, on lugs made of the Cu–Mg alloy 2024-T3 and of the Zn–Mg alloy 7075-T7351. All the specimens had sharp mechanical saw cuts at the critical section of the lug to act as crack starters. It had been observed that crack growth from artificial starters is more rapid than from natural ones (fatigue cracks) because these are irregularly shaped. A typical set of results is shown in Fig. 18.

A number of practically important conclusions were drawn from the investigation: the crack propagation life of an initially cracked lug is a small fraction of the fatigue life of an undamaged one. Undamaged 2024 alloy lugs showed larger spectrum fatigue lives than similar 7075 alloy lugs. However, lugs of both materials had similar crack propagation lives with a starter crack on one side of the hole. Measured crack propagation lives for both alloys fell in nearly common scatter bands for constant amplitude and spectrum loadings.

Theoretical estimates gave only rough crack propagation predictions under constant amplitude loading which were not all conservative. The authors suggest that, although good agreement was obtained with tests using a Miner's damage summation,[27] this should not be taken as general because of the limited number of specimens tested and the nature of the loading programs.

5 FRICTION GRIP BOLTED JOINTS

The transfer of load tangentially between the parts of a joint using mechanical fasteners can be through the fasteners, acting as dowels, or by friction between the connected parts induced by the clamping action of pre-tensioned fasteners. Although the existence of some clamping action caused by the shrinkage of hot driven rivets has long been recognised, the use of quantifiable frictional action in structural design has become common only in the last 40 years. Friction bolting provides a more rigid joint than ordinary bolting and is therefore useful where overall geometrical distortion is to be minimised. Under fluctuating loading, especially where the sign of the load changes, a very greatly improved fatigue life is obtained.

The primary design parameters for friction grip bolted joints are the slip coefficient for the faying (contact) surfaces of the joint and the pre-tension in the bolt which provides the clamping action. The bolts being normally fitted in holes with clearances of up to 2 mm transmit no load by direct transverse shear.

Before joint behaviour can be adequately explained, it is necessary to examine the actions between rough contacting surfaces and the manner in which the normal pressure between them, resulting from the bolt clamping force, is distributed.

5.1 The transfer of static load by interface friction

The mechanism of load transfer at a single contact between solid surfaces, subjected to normal and tangential forces, has been described by Mindlin,[28] who showed that interface shear stress (traction) is proportional to the local relative interface displacement. Beyond a limiting value determined by the coefficient of friction, irrecoverable displacement (slip) occurs with constant shear resistance.

The normal pressure (z) on the interface of two flat plates bolted together has been shown by Cullimore & Eckhart[29] to be distributed

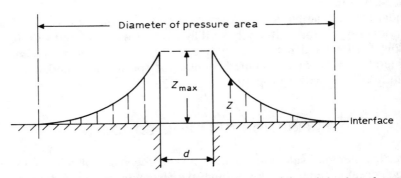

Fig. 19. Distribution of bolt clamping pressure (z) on joint interface.

over an annulus around the bolt hole of the order of twice the bolt diameter. The pressure is greatest at the edge of the hole, falling rapidly to zero at the periphery of the annulus (Fig. 19). The maximum pressure, which is a function of $\Sigma t/d_f$, where Σt is the total joint thickness, decreases as Σt increases.

In a joint the interface pressure, and thus the shear resistance, is zero at the periphery of the bolt pressure area, so that slip occurs there at the outset of loading and progresses inwards as the load increases. The displacements are small at this stage because of the restraint provided by the remainder of the bolt pressure area where the interface shear has not reached its limiting value. As the load is increased, the displacement increases more rapidly until, when the limiting shear condition is reached over the whole of the bolt pressure area, there is no further resistance to slip until the hole clearance is taken up and the bolt bears on the plate.

This behaviour is illustrated by the results of a slip test on a butt joint with a 25 mm thick aluminium centre plate having one row of two M22 bolts as shown in Fig. 20. The relative displacement between the plates was measured on a transverse cross-section through the centre of the leading bolt. The fastener flexibility relation, needed to calculate the fastener resistances in a multi-bolted joint with untensioned bolts, is of a similar form. When two macroscopically smooth surfaces are pressed together the normal force is resisted by the yielding of opposing surface asperities. An applied in-plane force will be resisted by shearing in these junctions. The resulting tangential displacement has been shown by Green[30] to be accompanied by an inward normal plastic displacement moving the surfaces closer to-

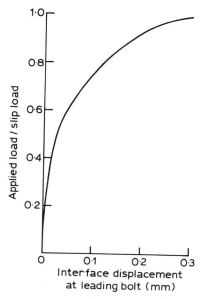

Fig. 20. Slip in a HSFG bolted aluminium joint. $\mu = 0.54$.

gether. In a joint, these normal displacements, by reducing the joint thickness, will shorten the bolt and so cause an irrecoverable loss of pre-tension. Higher bolt tensions and rougher surfaces increase the loss of pre-tension. Negative through-thickness strains accompany applied in-plane tensile forces. The loss of pre-tension from this elastic Poisson's ratio effect is, however, recoverable. These effects are illustrated in Fig. 21. In both cases, the loss of tension can be reduced by increasing the axial flexibility of the bolt. If the in-plane stress in the region of stress concentration at the edge of the hole increases to a value which, combined with the peak normal stress there, results in local yielding of the plate, the increased thinning rapidly reduces the bolt tension. Thus, although the frictional shear resistance of an interface is effectively constant for a particular surface; because of the variation of bolt tension the slip coefficient, which is calculated from the initial bolt pre-tension, will vary with the joint geometry and strength of the plate material. When selecting slip factors for design it is important to use either values obtained from tests on specimens of the same geometry as the actual joint, or to modify the nominal values from standard test joints.

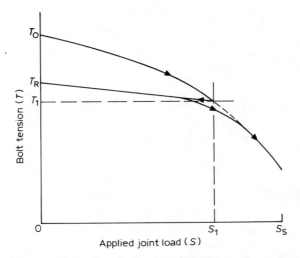

Fig. 21. Reduction of bolt tension with applied load in a shear joint. S_s, joint slip load; T_o, initial bolt tension; T_1, bolt tension at joint load S_1; T_R, bolt tension after unloading from S_1; $(T_R - T_1)$, elastic component of reduction of bolt tension at S_1.

These factors are illustrated by the results of tests on joints with grit blasted steel plates reported by Cullimore[31] shown in Table 1.

It will be noted that although increasing bolt pre-tension gives a higher slip load, because of the proportionally larger loss of bolt tension, there is a slight decrease in slip factor. At the smaller plate width some yield of the net section is indicated which accounts for the reduction of the slip factor. Also of note is the large loss of bolt tension on unloading after slip.

Table 1. Slip Tests on Double Cover Plate Butt Joints with two M20 Bolts

Bolt pre-tension (kN)	144		190	
Plate width/hole diameter	4	2·5	4	2·5
Slip load (kN) (S_s)	251	234	296	275
Slip factor	0·45	0·41	0·39	0·37
$S_s/Y_n{}^a$	0·53	1·07	0·63	1·25
Bolt tension at $0·9S_s$ (kN)	127	122	162	155
Bolt tension after unloading (kN)	97	98	141	105

^{a} Y_n = load to cause yield on net section.

The slip load will be increased by increasing the friction coefficient of the faying surfaces. Roughening the surface by shot blasting increases the slip resistance, partly by increasing the friction coefficient because of the removal of contaminants, and partly by providing some addition to the effective shear area. However, the improvement is partially offset by the increased loss of bolt tension. The rougher surface will also increase the creep relaxation of the bolt pre-tension, as will surface coatings such as metal spray. These effects may be mitigated by increasing the axial flexibility of the bolt, e.g. by using higher strength bolts of smaller diameter, and by reducing the peak normal interface pressure by increasing the joint thickness/bolt diameter ratio ($\sum t/d_f$).

Real contact between the plies is limited to the bolt pressure area which will, in most cases, be small in relation to the area of the joint interface. Initial lack of flatness of the plates may cause the bolt load to be transferred to areas away from the bolt pressure area. This will not affect the slip resistance unless the irregularity of the surface is such as to reduce these contact areas sufficiently to cause local yielding. In cases where the initial out-of-plane distortion of the plates is large, if the plates are stiff enough, a significant part of the bolt pre-load may be transferred to the opposite half of the joint and the slip resistance will be reduced.

5.1.1 Slip factors for aluminium alloy joints

Aluminium alloy rolled and extruded materials, as received, have smooth surfaces giving a coefficient of friction of the order of 0·10–0·15, which is too low for practical use with friction bolting. Cullimore[32] pointed out in 1959 that an increase in slip coefficient, sufficient to make the use of friction bolting practicable, could be made by grit blasting the faying surfaces. Later investigation by Valtinat[33] and Molina & Hacquart[34] produced data which were used in the first (1978) edition of the ECCS European Recommendations for Aluminium Alloy Structures (ERAAS). Three surface treatments were included: grit blasted, grit blasted with zinc silicate coating, and grit blasted with resin bonding. Values of the slip factor, μ, were given for three ranges of thickness (t) of the thinnest plate in the joint, varying from 0·2 for 3 mm < t < 6 mm up to 0·55 for resin bonded surfaces with t > 10 mm. Subsequently an extensive programme of tests done by Ramirez[35] enabled the earlier ERAAS slip factor recommendations to be revised. Up to ten results were obtained for

Table 2

Total joint thickness, $\sum t$ (mm)	$12 < \sum t < 18$	$18 < \sum t < 24$	$24 < \sum t < 30$	$\sum t > 30$
Characteristic slip factor (μ)	0·30	0·35	0·40	0·45

each specimen configuration, and a characteristic value of the slip factor (μ) calculated from

$$\mu = \mu_m - 1·28sd \qquad (27)$$

where μ_m was the arithmetical mean of the results and sd their standard deviation, implying a nominal probability of 10% of the slip factors being less than the characteristic value. The results given for 6061-T6 plate with lightly shot blasted surfaces (N10—Rugotest No.3B range) are shown in Table 2. It is to be noted that the data are for a single size of bolt (M16), so that the effect of variations of the important parameter ($\sum t/d_f$) are not included.

It was concluded (inter alia) that: zinc silicate and epoxy coatings gave poor resistance to long-term loading, light shot blasting was a satisfactory surface treatment, and that the slip factor was independent of the number of bolts in the joint. This last observation, although valid for the compact joints tested, would need to be modified for long joints for the reasons described previously for ordinary bolted joints.

5.1.2 Attainment and stability of the bolt pre-load

Effective friction bolting requires a high pre-load stress—commonly 90–100% of the 0·2% proof stress of the bolt material—such as can be provided by high tensile steel or titanium bolts. With aluminium joints the resulting high bearing stresses under the nut and the bolt head are likely to cause surface damage and local indentation if the strength of the bolt material exceeds about three times that of the plate material. This damage will cause loss of accuracy in the control of bolt pre-tension, increase bolt relaxation, and form sites for fatigue crack initiation. The damage can be eliminated by using thick steel 'washers' or having steel outer plies; the features of the latter arrangement will be referred to later.

Reliability of slip resistance calls for consistent bolt pre-tension. The commonly used methods of bolt tightening—torque control and part-turn—are known to produce wide variations of bolt tension. These variations can be reduced by using good new lubricated threads and, with the torque control method, by limiting the specified pre-load to 70% of the ultimate tensile strength of the bolt. Improved control of pre-tension may be obtained by the use of load indicating washers, yielding sensing torque wrenches, and devices which measure the extension of the bolt either directly or ultrasonically. The increased cost, in terms of hardware and assembly time, associated with using some equipment of the two latter types could be justified only where a very high degree of reliability is required.

On releasing the tightening action, the bolt tension will relax at a rate that will rapidly decrease with time, so that most of the loss of pre-tension will take place in the first 12 h after tightening. This effect is caused by creep in the highly stressed thread contact areas, and in the local areas of contact of the plies. The former will depend on the properties of the bolt material, and will increase with stress level; the latter on the roughness of the surfaces and any coating that has been applied. The consequent shortening of the bolt, being independent of its length, will have its most severe effect on reducing pre-tension at the shorter grip lengths.

In tests on a series of surfaces, Cullimore (loc. cit.[31]) showed that the relaxation after 12 h varied from as little as 5% for steel plates with untreated surfaces, up to 30% for surfaces which had been grit blasted and thickly sprayed with zinc. Two sample M22 bolts were tightened, each in a pack of plates of similar thickness to the joint in which they were to be used, and the tensions were continuously monitored from 1 min after tightening. After 7 days (Fig. 22), the relaxations observed were 1·5% in bolt A and 2·3% in bolt B. Re-tightening bolt B produced an immediately increased rate of relaxation which, however, after a lapse of about 10 weeks, had reduced to approximately that of bolt A. It was estimated that about 70% of the additional tension induced by the re-tightening would be retained, so providing a useful enhancement of the long-term slip resistance.

Variations in relaxation values quoted by different investigators may be attributed to differences in specimen geometry and surface treatments, uncertainty in defining the initial peak tension value, and variations due to drift in the monitoring equipment

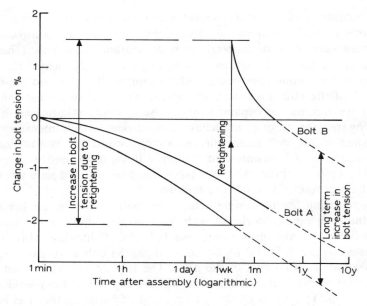

Fig. 22. Relaxation of tension in an M22 bolt.

5.2 HSFG bolted joints between aluminium and steel plates

The limitations to increasing the slip resistance of aluminium joints can be overcome by using steel outer plates in the butt joint; because their resistance to local stresses under the nut and bolt head is greater than that of aluminium plates, higher bolt stresses may be used to give a larger pre-tension for a given bolt size. A rough steel surface in contact with the aluminium might also be expected to enhance the interface shear resistance. These advantages may, however, be partially offset by the increased relaxation of bolt tension resulting from larger creep rates at the higher stresses, and by the thermal stresses induced by the differential expansion of the aluminium and steel. Furthermore, bi-metallic corrosion will be a problem in some environments.

The first two of these problems have been discussed by Cullimore & Millward[36] in relation to the development of a compact high-performance joint using aluminium and steel plates. Bolt tension relaxation tests were conducted in which bolt tensions were continuously monitored for up to 10 weeks. Superimposing a constant temperature cycle in these tests demonstrated that, after the initial period, the two effects could be considered independently.

5.2.1 Estimation of the stresses induced by changes in temperature

If a steel bolt is tightened in an aluminium joint, because the coefficient of thermal expansion of aluminium is approximately twice that of steel, a drop in temperature after assembly will result in a shortening of the bolt causing a loss of pre-tension. Conversely, a rise in temperature will extend the bolt and, if large enough to cause yield, will result in permanent loss of pre-tension on return to ambient temperatures. The method of calculating the change (Θ) in bolt force caused by a unit change in temperature is illustrated by reference to the simple assembly, consisting of an aluminium plate, thickness t_a and a steel plate of thickness t_s, shown in Fig. 23.

$$\text{Change in length of steel fastener} = (t_a + t_s)\theta_s + f_t\Theta \qquad (28)$$

$$\text{Change in thickness of the plates} = (t_a\theta_a + t_s\theta_s) - f_p\Theta \qquad (29)$$

where the coefficients of thermal expansion of aluminium and steel are θ_a and θ_s; the axial flexibility of the bolt is f_t and that of the plate assembly is f_p. Provided the plates remain in contact, these two changes in length are equal, so that

$$\Theta = t_a(\theta_a - \theta_s)/(f_t + f_p) \qquad (30)$$

The value of Θ, the tension/temperature coefficient, is immediately obtainable once the bolt and plate flexibilities are known.

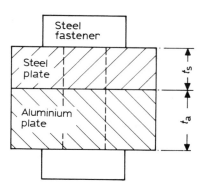

Fig. 23. Simple clamped assembly of aluminium and steel plates.

The plate flexibility was found by using a finite element analysis to calculate the displacement between the outer surfaces of the plates caused by a unit axial compressive force. Friction effects due to differential lateral displacements of the plates were not included in the model. It was found that, for the four combinations of plate and thickness examined, the area of plate effective in reacting bolt force was equivalent to a cylinder of diameter 4·0–4·7 times that of the bolt. Using a value of 4 × bolt diameter should give a conservative estimate for Θ.

The bolt flexibility (f_t) is the displacement between the inside faces of the nut and bolt head caused by unit axial tension force. It is made up of the extension of the plain part of the shank (L_1/EA_1), that of the threaded part of the shank in the grip length L_2/EA_2 (see Fig. 24), and the displacement due to the nut flexibility (f_n). The latter comprises deformations of the bolt head, nut, and those of the mating threads in the nut. The bolt flexibility is therefore:

$$f_t = L_1/EA_1 + L_2/EA_2 + f_n \tag{31}$$

Because of the extreme complexity of the modelling of the mating threads, direct experimental measurement of the overall bolt flexibility

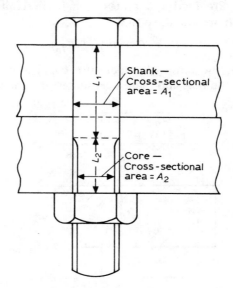

Fig. 24. Bolt geometry.

is preferred to finite element analysis. The nut flexibility is then obtained by subtracting the shank flexibilities calculated from the known bolt dimensions (eqn (31)) from the measured overall value, f_t. The nut flexibility can be expected to increase with increasing thread clearance and with decreasing proof strength of the bolt material.

5.2.2 Application to an actual design for static loading

Tests showed that f_n was effectively constant for a wide range of plain and threaded shank lengths in a given size of bolt. An average measured value of f_n for M22 was 0·87 μm/kN, giving the flexibility of 1·86 μm/kN for the bolt with a grip length of 75 mm (L_1 = 65 mm, L_2 = 10 mm). Within the limited range of sizes examined, f_n was always found to make a similarly substantial contribution to the overall flexibility. Thus omitting the nut flexibility, as is commonly the practice, will generally lead to a significant over-estimation of the tension/temperature coefficient, Θ. If the plate flexibility is negligible, eqn (30) reduces to

$$\Theta = t_a(\theta_a - \theta_s)/f_t \qquad (32)$$

In this case, where f_p = 0·256 μm/kN and f_t = 1·86 μm/kN, neglecting f_p over-estimates Θ by about 14%.

Good agreement was obtained between calculated and measured values of Θ for single bolted assemblies. With closely spaced groups of bolts the restraint of out-of-plane flexure of the plates was expected to reduce the plate flexibility and less good agreement was obtained.

The joint which was developed in the study used M22 grade 12·9 steel bolts (nominal proof tension 283 kN); the surfaces of the aluminium plates were untreated, the grip being obtained by blasting the steel surfaces with a coarse grit (BS2451 Grade 66). The mean value of the slip factor achieved in short duration tests was 0·56 with a standard deviation 0·015 (2·68%).

The calculated value of the tension/temperature coefficient was 0·161 kN/°C, so that a drop in temperature of 50°C below that at which the connection was assembled would be expected to reduce the bolt pretension by 8 kN (1·84%); this recoverable loss will be largely lost in the scatter of the values of slip resistance (2·68%).

The long-term creep loss of pre-tension was estimated to be 7·5 kN, which must be deducted from the characteristic pre-tension in estimating the long-term slip resistance. A rise in temperature shortly after tightening, before significant relaxation of the maximum torqued-tension load has occurred, will cause further inelastic extension and an irrecoverable loss of pre-tension which must be added to the long-term creep loss.

In the joint for which the above values have been calculated, the thickness of the aluminium plate was 23 mm and the total thickness of the steel plates was 52 mm. It would be more usual for the aluminium plate to be thicker than the steel. If the thicknesses were reversed, the tension/temperature coefficient would be roughly doubled.

5.2.3 Bonded HSFG bolted joints

The partition of load between the bolts and the adhesive in a bonded joint will depend on their relative shear flexibilities. Under service loading (i.e. below slip), an increase in joint stiffness will be obtained which is likely to be significant for large bolt flexibilities only, e.g. with low slip factors. Similar factors affect the contribution of the adhesive to the ultimate strength of the joint. Pre-tensioned bolts at the edges of the bonded lap should provide resistance to peeling.

In a small number of tests made by Wise[37] on butt joints in 6083-T6 plate with M12 HSFG bolts, bonding the interfaces reduced the overall deflection at service loads by about 10%. A significant reduction in deflection of joints with untensioned bolts was obtained by filling the clearance between the bolt and the hole in the plate with adhesive.

5.3 Fatigue resistance

In fatigue tests on single bolted joints, Perrett[14] demonstrated that pre-tensioning the bolts effected a large increase in fatigue strength compared with joints having plain, untensioned bolts. In the plain bolted joints (Fig. 25), failure occurred from cracks originating at the inside surface of the bolt hole. In the clamped joints, cracks initiating in the interface fretting area propagated inwards towards the hole and outwards towards the edges of the plate. Fretting occurs in areas of contact where there is to-and-fro relative movement (reversed slip) between the plies. The dark band in Fig. 26 is the fretting area around the leading bolt hole in an aluminium joint plate.

Edwards & Ryman[38] studied the effects of fretting on axially loaded specimens of 2014-T6 bar. The fretting was produced by the usual bridge device clamped to the specimen which provided two contacts of essentially uniform pressure. In the case shown in Fig. 27, fretting

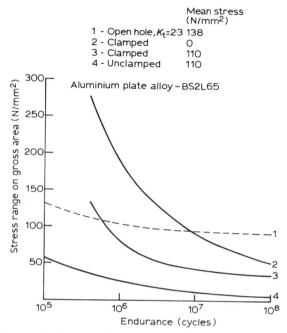

Fig. 25. Effect on fatigue endurance of clamping in a single bolt butt joint.

reduces the fatigue strength by a factor of about three, the damage becoming less as the fatigue stress range is reduced. Clamping together the plates of a joint, by removing the failure crack initiation site from the high stress fretting contact area in the wall of the bolt hole to the

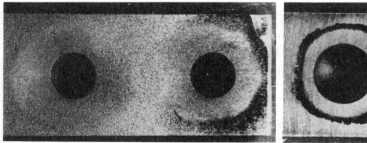

Fig. 26. Fretting damage in HSFG bolted aluminium joints.

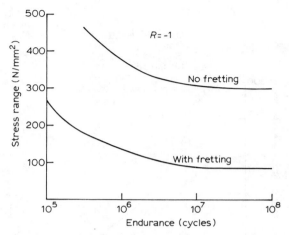

Fig. 27. Effect of fretting on aluminium alloy 2014-T6 extruded bar in axial fatigue loading.

lower stressed fretting zone in the bolt pressure area of the plate, increases the fatigue strength.

Several investigators have developed fracture mechanics models of fretting fatigue to calculate crack growth under the conditions of the laboratory fretting bridge fatigue test. Nix & Lindley[39] report satisfactory agreement with the results of tests on aluminium alloy 2014A specimens in contact with steel fretting pads. The extension of these single contact models to the very complex stress conditions of a joint interface does not yet appear to have proved to be practicable.

5.3.1 Fretting in HSFG joints
Fretting damage is governed primarily by:

(i) The distribution of bolt clamping pressure over the interfaces, which is a function of $\sum t/d_f$ and $\sum t/t_c$ where $\sum t$ is the total joint thickness and t_c the thickness of the cover plate.

(ii) The relative displacement between the interfaces in the bolt pressure area, which is a function of the maximum pressure, the nature of the surfaces and the traction stresses.

Other factors affecting the fatigue strength are the number and disposition of bolts in the joint, their edge and end distances, and the basic fatigue properties of the material of the plies.

5.3.2 The estimation of fretting displacements in HSFG bolted joints

The tangential force, or traction (Y), resisting movement between a pair of opposing interface elements in a joint is related to the relative displacement (δ) by the traction coefficient (λ), defined here as the tangential force transmitted between two elements of unit length per unit relative displacement, so that

$$Y = \lambda \delta \tag{33}$$

For two such elements, subjected to a uniformly distributed normal force Z, the limiting value of the traction is

$$Y_u = \mu Z \tag{34}$$

where μ is the slip (friction) coefficient for the interface surfaces. These simple traction/displacement relationships are illustrated in Fig. 28.

The generation of fretting displacements is demonstrated by the traction/displacement relationship (Fig. 29) for an interface element in a joint subjected to an external loading cycle $0-S_{max}-S_{min}-S_{max}$, where $S_{max} > S_{min} > 0$.

During the increase of the load from zero the displacement at the element reaches a value generating limiting traction (B in Fig. 29). Slipping then starts and the element offers no further resistance to the

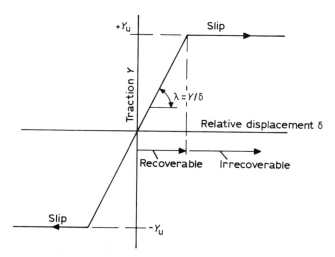

Fig. 28. Traction/displacement relation for a single pair of elements.

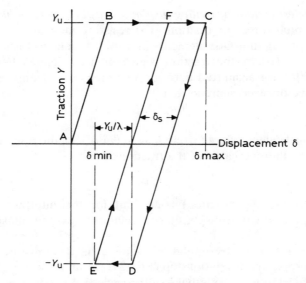

Fig. 29. Traction/displacement for an element in the joint interface.

increase of displacement up to $\delta_{max}(C)$ when the external load is S_{max}. The increase of load to S_{max} is resisted by other elements where the normal force is higher and, consequently, the resistance to slip is greater.

As the load is reduced from S_{max} the displacement decreases linearly until limiting traction is reached in the opposite direction (D). Further reduction of the displacement results in reversed slip of the element (see Fig. 28) until the lower load S_{min} is reached at a displacement $\delta_{min}(E)$. Re-loading causes a linear increase in displacement until limiting traction is again reached (F) and slipping at the element occurs until the maximum displacement, δ_{max}, is reached again at the load $S_{max}(C)$. In subsequent loading cycles the traction/displacement relation follows the closed loop CDEFC and the element suffers a to-and-fro slipping displacement δ_s, where

$$\delta_s = \delta_{max} - \delta_{min} - 2Y_u/\lambda \tag{35}$$

the amplitude of fretting for the element.

An appropriate analysis using a greatly simplified joint model consisting of two parallel line elements was developed by Cullimore & Millward[40] to estimate relative interface displacements, fretting amplitudes and traction, to provide a guide for fatigue test specimen joint design. A typical traction distribution in the first loading cycle is shown in Fig. 30 for a two-bolt butt joint where the axial stiffness of the outer plates is greater than that of the centre plate. When the upper load in the cycle, S_{max}, is reached, slip has occurred over a length OA of the joint in the first bolt pressure area, and over a length BC in the second. As the load falls to the lower value, S_{min}, negative limiting traction is reached, causing reverse slip, from O to D and from E to C. During subsequent loading cycles, to-and-fro slipping causes fretting over the lengths OD and EC. Direct measurement of the traction coefficient is not practicable, but it was found that using a value of $\lambda = 1.0$ MN/mm gave fretting zone sizes similar to those observed in the two-bolt joints, with grit blasted faying surfaces used in the tests described later. A parametric study of an 11-bolt joint showed that, for a value of $\mu = 0.6$, variations of λ from 0.5 to 5.0 MN/mm caused negligible changes in fretting displacement. Fretting displacement was, however, affected by changes in the coefficient of friction. For $\lambda = 1.0$ MN/mm the maximum fretting displacement was doubled by reducing μ from 0.80 to 0.45.

These estimations of fretting displacements and tractions, being based on simplified representation of the very complex conditions at

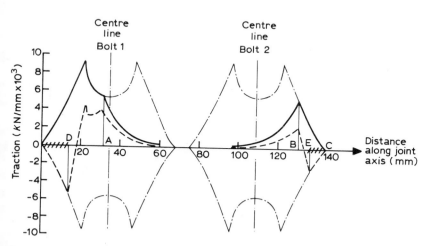

Fig. 30. Traction in a two-bolt joint. Loading cycle: $0 \rightarrow S_{max} \rightarrow S_{min} \rightarrow S_{max}$. Traction distributions at loads: S_{max}——, S_{min}————, Overall slip —·—·—·. Extent of fretted zone on interface ++++++.

the joint interface, cannot be expected to give precise quantitative solutions. They are, however, useful for comparing values of these parameters for various joint configurations.

5.3.3 Fatigue tests on two-bolt aluminium/steel HSFG bolted joints

Fluctuating axial tension ($R = 0$) fatigue tests on two-bolt double cover plate butt joints were carried out as part of the investigation referred to in the previous section. The specimens were designed for fatigue to take place in the 7000 series Al–Zn–Mg alloy centre plate (minimum $\sigma_{0.2} = 330\,\text{N/mm}^2$). The cover plates were BS EN26X steel plates (minimum $\sigma_y = 950\,\text{N/mm}^2$) whose faying surfaces were blasted with coarse grit to increase the slip factor. The bolts were 12·9 grade M22, with a pre-tension of 283 kN. The results (Fig. 31) lie between the σ_r–N curves for the plain and notched ($K_t = 2\cdot33$) plate material, tending to approach and cross the latter as the stress range decreased. Fretting cracks for the middle and lower stress ranges originated in the bolt pressure area forward of the leading bolt, as suggested by the analysis. At the higher stress range where slip had allowed a larger proportion of the load to be transferred to the minimum section in some specimens, cracks initiated at the bolt hole.

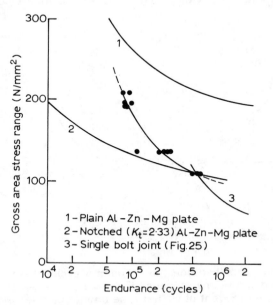

Fig. 31. σ_r–N curve for two-bolt aluminium/steel HSFG bolted joint.

The tests established the need to provide adequate edge and end distances for the bolts. Failure to do this, and allowing the bolt pressure area to be curtailed, resulted in significant reductions of endurance. Increasing the steel cover plate thickness, and so increasing the out-of-plane bending stiffness, resulted in fretting at their outer edges and a reduction in endurance. Pre-loading the joint was found to have a beneficial effect on the fatigue resistance by inducing residual compressive stresses at critical areas of the interface.

5.3.4 Multi-bolted joints

It was shown, when dealing with static loading, that the end fasteners of a row resist more of the load than those towards the centre, and that their slip resistance may be overcome at a load significantly less than that causing overall slipping. In such a case, traction in the pressure area of the end bolt having reached its limiting value, further increases in applied load are reacted directly by increases of stress on the cross-section at the bolt hole. Eventually the stress may reach a level at which, in fatigue loading, the stress concentration effect of the hole causes fatigue cracks to initiate there. The fatigue life for this form of failure is likely to be less than that when cracks are initiated by fretting of the plate in front of the bolt hole. Furthermore, increasing the length of a joint increases the relative strains between the plates, and hence the fretting displacements which, up to a limiting value, also reduce the fatigue life. The aim of design against fatigue should therefore be to make joints as compact as possible.

5.3.5 Fatigue testing of full-size joints

It frequently happens that the maximum loads required for the fatigue testing of a full-size joint exceed the capacity of the available testing equipment. Fabricating and testing a large enough number of specimens to obtain acceptably reliable data may frequently be prohibitively costly. In either case, the use of models will have to be considered. The option of retaining geometrical similarity by reducing scale should be rejected because, particularly with mechanical fasteners, this is known to give erroneous results in fatigue. A reasonable compromise can be obtained by using fewer bolts of the full size, making the applied load per bolt as nearly as possible the same as in the prototype and adjusting the cross-section to give the appropriate stress levels.

With HSFG bolted joints, additional conditions of similarity for interface fretting and traction have to be satisfied. Fretting damage is governed by the clamping pressure, the surface traction stresses and the relative interface displacements. Using the prototype bolt pretension and the same plate thicknesses will ensure the correct peak clamping pressure and distribution. Fatigue failure will generally occur in the plate at, or in front of, the first row of bolts. The required fretting displacement in this area may be achieved by varying the distance between the first and second bolts. For similarity of traction conditions the in-plane stresses must be the same at any value of the ratio of applied load/slip load for model and prototype—a condition which is difficult to satisfy. In most cases it will not be possible to satisfy all the conditions for similarity, and compromises have to be made which result in a loss of accuracy which may vary with the applied load. The result will improve as the number of bolts in the model approaches that in the prototype.

REFERENCES

1. Bulson, P. S. & Cullimore, M. S. G., Design problems in welded aluminium structures. Fourth International Conference on Aluminium Weldments, The Japan Light Metal Welding and Construction Association, Tokyo, 1988, 5-94-100.
2. Fisher, J. W. & Struik, J. H. A., *Guide to Design Criteria for Bolted and Riveted Joints*. John Wiley, New York, 1974, p. 53.
3. Marsh, C., Tear-out failure of bolt groups. *Proc. Am. Soc. CE*, **105** (ST10) (Oct. 1979) 2122–5.
4. ESDU International, *Elastic Stresses in Single Lap Joints Under Tension or Compression*. Data Item 67008, London 1967.
5. Batho, C., The partition of load in riveted joints. *J. Franklin Inst.*, **182**(5) (1916) 553–604.
6. Francis, A. J., *Investigation on Aluminium Alloy Riveted Joints Under Static Loading, Research Engineering Structures, Supplement*. Butterworths, London, 1949, 187–215.
7. Flint, A. R., Analysis of behaviour of riveted joints in aluminium alloy ships' plating. *North East Coast Inst, Engrs. and Ship Builders Trans.*, **72**(3) (1955) 83–122.
8. ESDU International, *Flexibility of a Single Bolt Shear Joint*. Data Items 85034 and 85035, London, 1985.
9. Fisher, J. W. & Rumpf, J. L., Analysis of bolted butt joints. *J. Struct. Div. ASCE*, **91** (ST5) (October 1965) 181–203.
10. Francis, J. W., *The Behaviour of Aluminium Alloy Riveted Joints*, Research Report No. 15. The Aluminium Development Association, London, March 1953.

11. Crawford, S. F. & Kulak, G. L., Eccentrically loaded bolted connections. *J. Struct. Div. ASCE*, **97** (ST3) (March 1971) 765–83.
12. Kamtekar, A. G. & Wittrick, W. H., Limit analysis of fastener group under eccentric load. *Int. J. Num. Meth. Engng*, **20**(6) (June 1984) 131–51.
13. Marsh, C., Discussion on Brandt, G. D., Rapid determination of ultimate strength of eccentrically loaded bolt groups. *Engng. J. Am. Inst. Steel Constr.*, **19**(2) (1982) 94–100, 214–5.
14. Perrett, B. H. E., *Fatigue Endurance of Structural Elements in Various Materials Under Constant and Variable Amplitude Loadings*. RAE TR77162, HMSO, London, 1977.
15. Heywood, R. B., *Correlated Fatigue Data for Aircraft Structural Joints*. ARC Current Paper 227, 1956.
16. McLester, R., Fatigue strength of welded and riveted joints in aluminium. *Symposium on Aluminium in Structural Engineering*, Aluminium Federation, London, 1964, pp. 9–19.
17. ESDU International, *Endurance of Riveted Lap Joints (Aluminium Alloy Sheet and Rivets)*. Data Item 79031, London, 1979.
18. Welter, G., Fatigue of spot welds. *Welding Int.* **34** (1953) 145s.
19. Cox, H. L. & Brown, A. F. C., Stresses around pins in holes. *Aero. Quart.*, **XV** (4) (Nov. 1964) 357–72.
20. Blakeborough, A. B., Department of Civil Engineering, University of Bristol. Private communication, August 1988.
21. Cullimore, M. S. G. & Mason, P. J., Fatigue and fracture investigation carried out on the Clifton suspension bridge. *Proc. Inst. Civ. Engrs*, **84** (Part 1) (April 1988) 309–29.
22. Webber, D., *Proposal for a change in the method of pinned-joint design*. MVEE Branch Note EE/BR—105-83, HMSO, London, 1983.
23. ESDU International, *Endurance of Aluminium Alloy Lugs with Nominally Push-Fit Pins (Tensile Mean Stress)*. Data Item 80007, London, 1980.
24. Rooke, D. P. & Cartwright, D. J., *Compendium of Stress Intensity Factors*. HMSO, London, 1976.
25. Jaske, E., Fedderson, C. E., Davies, K. B. & Rice, R. C., *Analysis of Fatigue, Fatigue Propagation and Fracture Data*. NADA CR-132332, 1973.
26. Buch, A. & Berkovits, A., *Fatigue crack propagation in Al-alloy lugs under constant amplitude and maneuver spectrum loading*. Technion-Haifa, TAE Rpt. No. 497, 1983.
27. Miner, M. A., Cumulative damage in fatigue. *J. Appl. Mech.*, **12** (1945) A159.
28. Mindlin, R. D., Compliance of elastic bodies in contact. *J. Appl. Mech.*, **16** (1949) 259–268.
29. Cullimore, M. S. G. & Eckhart, J. B., The distribution of clamping pressure in friction grip bolted joints. *Struct. Engng*, **52**(5) (1974) 129–31.
30. Green, A. P., The plastic yielding of metal junctions due to combined shear and pressure. *J. Mech. Phys. Solids*, **2** (1954) 197–211.
31. Cullimore, M. S. G., *Bolted Connections—Research Affecting Current Design Practice. The Design of Steel Bridges*. Granada, London, 1981, pp. 421–32.

32. Cullimore, M. S. G., Tests on an aluminium alloy joint with pre-tensioned bolts. *Proc. Jubilee Symposium on High Strength Bolts,* Institution Structural Engineers, London, June 1959.
33. Valtinat, G., Untersuchung zur Festlegung zulassiger Spannungen und Krafte bei Niet-Bolzen-und HV-Verbindungen aus Aluminium legierungen. *Aluminium,* **47** (1971) 735–40.
34. Molina, C. & Hacquart, P., *Emploi des Boulons a Haute Resistance a Serrage Controle.* Rpt. No. 1459, Centre Technique de l'Aluminium Pechiney, Paris, 1976.
35. Ramirez, J. L., *Aluminium Structural Connections: Conventional Slip Factors in Friction Grip Joints. Aluminium Structures: Advances, Design and Construction.* Elsevier Applied Science, London, 1987, pp. 115–25.
36. Cullimore, M. S. G. & Millward, C. P., Special features in the design of steel/aluminium HSFG bolted joints. Steel structures. Recent research advances and their applications to design. *Proc. Int. Conf. Budva, Yugoslavia,* Civil Engineering Faculty, Belgrade University, Belgrade, 1986, pp. 317–26.
37. Wise, K. P., *A Comparison of Performances of Common Steel and Aluminium Bolted Connections Modified Using Adhesives.* Project Report, Department of Civil Engineering, Teesside Polytechnic, June, 1980.
38. Edwards, P. R. & Ryman, R. J., Studies in fretting fatigue under service loading conditions. *Proc. 8th ICAF Symposium,* Lausanne, June, 1975.
39. Nix, K. J. & Lindley, T. C., The application of fracture mechanics to fretting fatigue. *Fatigue Fract. Engng Mater. Struct.,* **8**(2) (1985) 143–60.
40. Cullimore, M. S. G. & Millward, C. P., *Development of a Fatigue-Resistant Aluminium HSFG Bolted Joint. Aluminium Structures: Advances, Design and Construction.* Elsevier Applied Science, London, 1987, pp. 162–70.

Index

Aircraft design, 2
  fatigue and, 14
Aluminium
  properties of, 1–5
  uses for, 1–3
Aluminium alloy joint slip factors, 349–50
Aluminium–steel HSFG bolted joints, 352–6
  fatigue, 362–3
  temperature and, 353–5
Amplitude loading, 340
ANSYS program, 337
Aramid aluminium alloys, 5
Axial loading, in-plane, 321–8
Axial tension, 318–19

'Ballising', 340
Bars
  definition of, 35–6
  geometrical imperfections, 36, 43–7
  mechanical imperfections, 36, 47–61
  stress–strain, 36–43
Beam buckling, lateral–torsional, 193–217
Beam codification, 179–80
  see also Codification; BS8118; CP118
Beam column buckling, 169–76
  experimental results, 172

Beam column buckling—*contd.*
  interaction domains, 170–1
  physical behaviour of, 169–70
  simulation results, 173–6
Bending, members in, 70–131
  codification, 127–31
  cross-section ultimate behaviour, 75–92
  flexural torsional buckling, 109–27
  plastic behaviour, statically undetermined girders, 92–109
Bolt clamping pressure, 345–9
Bolt flexibility, and temperature, 353–5
Bolt pre-load, 350–1
Bolt pre-load stability, 350–1
Bolt joints, 332–5. 345–64
  aluminium–steel, 362–3
  bonded HSFG, 356
  multi-, 363
Bridges, 4
BS8118, 8–13, 15–17, 24, 26–7, 69, 184–5, 219–56
  experimental and, 248–51
  overview of, 221–3
  standard CP118 and, 246–8
BS design curve, 202–5
Buckling
  beam column, 169–76
  column, 143–69
  lateral–torsional beams, 193–217

Buckling check codification, 129–31
Buckling curves, with members in compression, 131–3
Butt welds, 257

CAN3–5 157–N83, 180
Canadian code, 180–1
Cherry tests, 214
Clark–Massonnet formula, 172, 179
Clark–Rolf tests, 111, 114
Clarke–Jombock tests, 111, 113, 213
Code BS8118. *See* BS8118; Codification; CP118
Code CP118. *See* CP118; Codification; BS8118
Codification, 69–70, 127–31, 176–89
  beams, 179–80
  buckling check, 129–31
  Canada, 180–1
  columns, 176–9
  French, 181–2
  German, 183–4
  Great Britain, 184–5
    see also BS8118
  Italy, 185, 189
  plastic design, 127
  see also BS8118; CP118
Coefficients for actions, 5–6
Column buckling, 143–69
  curve evaluation, 152–61
  ECCS and, 145
  extruded members, 152–5
  extruded profiles, 150–1
  physical behaviour of, 143–4
  simulation results, 161–9
  welded members, 155–61
  welded profiles, 151–2
Column buckling curve evaluation, 152–61
  extruded members, 152–5
  welded members, 155–61
Column codification, 176–89
  see also Codification; BS8118; CP118
Compression, members in, 11–13, 131–89
  codification, 177–89

Compression, members in—*contd.*
  column buckling, 143–69
  ultimate behaviour, cross-sections, 133–43
Connected ply resistance, 318–29
Conventional elastic moments, 75–8
  extruded members, 75–8
  welded members, 78
Conventional reduced elastic moments, 133
CP118, 15, 246–8
  BS8188, and, 246–8
  see also Codification; BS8118
Crack propagation, of lugs, 341–5
Crack tip stress intensity factor, 18–19
Crawford–Kulak equation, 331
Cross-section dimension variation, 44–7
Cross-section, ultimate behaviour, 75–92, 133–43
  conventional elastic moment, 75–8
  conventional reduced elastic moment, 133
  fully plastic moment, 81–92
  plastic adaptation moment, 78–81
  reduced plastic moment, 134–43
Cross-section variation, 44–7

Dangelmaier method, 150
Design, 1–34
  curve positioning, 196
  fillet welds, 293–7
  joint fasteners, 328
  welded connections, 261–4
DIANA, 304
DIN4113, 182
Double T columns, 150–2
Double T profiles, 47–9
Ductility, of connections, 67–9
Dumont–Hill tests, 111–12
Dutheil method, 181
Dwight, J. B., 7–8

Eccentric loading, 328–32
ECCS committee, 145

Elastic analysis, 328–9
  in-plane eccentric loading, 328–30
Elastic distribution, fastener
  resistance, 321–4
Elastic moments, 133
Elasto-plastic fracture mechanics, 22
Elemental strength, 64–7
End bearing, and stiffener
  requirements, 241
ERAAS code, 180–9, 249–50
  see also Codification; BS8118;
  CP118
Eurocode, 4–6, 28
European Convention for
  Constitutional Steelwork, 5
Extruded members, 75–8, 81–90,
  92–103, 152–5
Extruded profiles, 47–50, 55, 150–1

Faella–Mazzolani method, 146–7
Fastener flexibility, 324–5
Fastener groups, 321–5
Fastener joints, 313–6
  friction group bolted, 345–64
  group, 321–5
  pinned, 335–45
  single in-plane loading, 316–21
Fatigue, 13–7
Fatigue crack propagation, lugs,
  341–5
Fatigue loading, 332–5, 339–45
  bolted joints, without pretention,
    332–4
  lug resistance, 340–5
  riveted joints, 334–5
  safe life design, lugs, 341
  spot-welded joints, 335
Fatigue resistance, 356–64
  full-size joints tests, 363–4
  HSFG joints, fretting, 358–62
  lugs, 340
  multi-bolt joints, 363
  two-bolt aluminium-steel HSFG-
    bolted joints, 362–3
Fillet welds, 257–61, 279–300
  design, 293–300
  parameters, 279–81

Fillet welds—*contd.*
  strength, 291–3
  test evaluation, 296–7, 306–10
  test results, 289–90
  test specimens, 281–7
  testing, 287–9, 304
  transverse/longitudinal, 257
  ultimate strength, 291–3
  weld areas, 290–1
Flanged beams, 209–11
Flexural torsional buckling, 109–27,
  172
  physical behaviour, 109–15
  simulation results, 115–27
Form $\varepsilon = \varepsilon(\sigma)$, 38–43
Form $\sigma = \sigma(\varepsilon)$, 38
Forman law, 342
Fracture mechanics, 17–24
French code, 181–2
Fretting, HSFG joint, 358
  damage to, 358
  displacement, 359–62
Frey–Massonnet method, 148
Friction group bolted joints, 345–64
Full-size joint fatigue testing, 363–4
Fully plastic moment, 81–90
  extruded member, 81–90
  welded members, 90–2

Geometrical imperfections, 36, 43–7
  cross-section dimension variation,
    44–7
  out-of-straightness, 43–4
German code, 182–4
Girders
  combined shear/bending, 244–6
  plastic behaviour of, 92–109
Grip bolted joints, 345–64
Group fasteners, 321–35
  in-plane axial loading, 321–8

HAZ. *See* Heat-affected zones
Heat-affected zones (HAZ), 7–9
  softening, 223–36
  transversely stiffened webs with,
    234–6

Heat-affected zones (HAZ)—*contd.*
  unstiffened webs with, 225–6
  unstiffened webs without, 226–34
  welded connections, 256–7
Hill–Clarke formula, 170
Hollow section welding, 302–4
HSFG bolted joints, 352–6
  fretting and, 358
  static load and, 355–6
  temperature and, 353–5

Industrial bars. *See* Bars
In-plane axial loading, 321–8
  design factors, 328
  elastic distribution, fastener resistance, 321–4
  elastic flexibility, 324–5
  ply resistance, 326–8
  ultimate strength, 325–6
In-plane eccentric loading, 328–32
  elastic analysis, 328–30
  ultimate strength analysis, 330–2
In-plane loading, 316–21
Italian code, 185, 189

Joints, and mechanical fasteners, 313–6
  fastener groups, 321–35
  friction group bolted, 345–64
  pinned, 335–45
  single fastener, in-plane loading, 316–21

Klöppel–Bärsch tests, 111, 114

Lap joints, 319–21
Large openings, in webs, 243–4
Lateral–torsional beam buckling, 193–217
  BS design curve, 202–5
  non-uniform moment, 205–9
  slender cross-sections, 211–14
  UK design procedure, 200–2
  unequal flanged beams, 209–11

Linear-elastic fracture mechanics, 19
Loading, constant amplitude, 340
Loading, and fasteners
  fatigue, 332–5, 339–45
  in-plane axial, 321–8
  in-plane eccentric, 328–32
  static, 336–9, 345–51, 355–6
Longitudinal fillet weld, 257
Longitudinal stiffness, 242–3
Longitudinally stiffened webs, in shear, 236–9
Lugs, 340–5
  crack propagation, 341–5
  resistance, 340
  safe life design, 341

Marsh method, 332
Mechanical fasteners, 313–66
Mechanical imperfections, 47–61
  inhomogeneous mechanical properties, 55–62
  residual stress, 47–55
Medium Girder Bridge, 4
Merchant–Rankine beam design, 195
m-factors, 208–9
Miner's damage summation, 345
Moment/shear interaction, 244–5
Moments
  non-uniform, 205–9
  reduced elastic, 133
  reduced plastic, 134–43
Monosymmetric I-beams, 210
Multi-bolt joint fatigue resistance, 363

Non-destructive testing, 23
Non-uniform moments, 205–9

Ogle, M., 16–7
Out-of-straightness, 43–4, 197–200

Parameter choosing, welded connections, 264–5, 301–2
  fillet welds, 279–81

Paris law, 22, 342
Perry curve, 11, 201
Perry–Robertson approach, 195
'Picture frame' mechanism, 228
Piecewise idealization, and stress–strain, 36–8
Pinned joints, 335–45
  design, 338–9
  fatigue loading, 339–45
  static loading, 336–9
  stress in, 336–8
Plane buckling, 172
Plastic adaptation moment, 78–81
Plastic behaviour, statically undetermined girders, 92–109
  extruded members, 92
  welded members, 103–9
Plastic design codification, 127
Plastic moments, 78–90, 134–43
  adaptation, 78–81
  fully, 81–90
Plate girders, 219–52
Ply resistance
  in-plane axial loading, 326–8
  mechanical fasteners, and, 326–7
Poisson ratio effect, 347
Pratt truss, 226–7
Probabilistic fracture mechanics, 23–4

Quality assurance, 25–7

Ramberg–Osgood law, 38–40, 42–3, 77, 82, 86, 90, 116, 118, 124, 134, 148, 161, 164, 175
Reduced plastic moment, 134–43
Règles AL (France), 181–2
Residual stress, and mechanical imperfections, 47–55
  extruded profile, 47–50
  welded profiles, 50–5
Riveted joints, 334–5

Safe life fatigue, lugs, 341
Safety, 5–6

Safety coefficients for action, 5–6
Shear webs, 219–52
Serviceability, 5–6
Single fastener joints, 316–21
  axial tension/shear, 318
  bearing of, 318
  connected ply resistance, 318–19
  design stress, 317
  lap joints, 319–21
  resistance, 317–21
  transverse shear, 317
Slender cross-section, 211–14
Slenderness, 131–3, 235
  ratio, 10–13
Slip factors, 349–50
Slip tests, 347–9
Spot welded joints, 335
Static load, 336–9
  interface friction, 345–9
Static loading, 355–6
Statically undetermined girders, 92–109
Stress–strain relationship, 36–43
  continuous models of form $\varepsilon = \varepsilon(\sigma)$, 38–43
  continuous models of form $\sigma = \sigma(\varepsilon)$, 38
  piecewise idealization, 36–8
Stiffened webs, transversely in shear, 226–36
  free from HAZ softening, 226–34
  with HAZ softening, 234–6
Stiffener requirement, 239–43
  end bearing, 241
  longitudinal, 242–3
  transverse, 239–40
  transverse bearing, 241
Stress, and temperature, 353–5
Stress, residual, 47–55
Sutter classes, 40–1

Tangential force, 359–62
Temperature, and aluminium–steel HSFG bolted joints, 352–6
Tension, members in, 9–10, 62–70
  codification, 69–70
  ductility, of connections, 67–9
  element strength, 64–7

Testing, 25–7
Torsional buckling, 109–27
Torsional beam buckling, 193–217
Traction/displacement, 359–62
Transverse bearing, and stiffener requirements, 241
Transverse fillet weld, 257
Transverse shear, 317
Transverse stiffeners, 239
Trilinear σ–ε law, 182–3

Ultimate behaviour, of cross sections, 133–43
Ultimate strength, fillet welds, 291–3, 298, 299
Unequal flanged beams, 209–11
UNI8643, 185, 189
Unstiffened webs, in shear, 223–6
  free from HAZ softening, 223–5
  with HAZ softening, 225–6

Valtinat–Miller method, 146, 150

Web openings, large, 243–4
Webs, 73–74, 223–6

Weld area, 290–1
Weld design, 257–61
  butt, 257
  fillet, 257–61, 279–300
Welded connections, 253–311
  aluminium alloys, 255–6
  butt welds, 257
  design of, 261–4
  fillet welds, 293–7
  HAZ, 256–7
  hollow sections, 302–4
  parameters, 264–5, 301–2
  research, 254–5
  results evaluation, 273–7
  state of the art, 255–64
  test results, 268–73
  test specimens, 265
  weld design, 257–61
Welded joints, 24–5
Welded members, 78, 90–2, 103–9, 155–61
Welded profiles, 50–61, 151–2

X-type specimen, 304–8

Young's modulus, 36, 38, 144